AF615695

# THE FIRST 49 PERSONALITIES

IN THE HONOR GALLERY OF THE AHA'S HEREFORD HERITAGE HALL

# THE FIRST

# PERSONALITIES

## IN THE HONOR GALLERY OF THE AHA'S HEREFORD HERITAGE HALL

**By DONALD R. ORNDUFF**

**With an Introduction by B.C. Snidow**

*A Series of Biographical Sketches Dealing With the Careers of Men Who Contributed Extraordinarily to the Hereford Breed and Its History in Recognition of the American Hereford Association's 100 Years—1881-1981*

**CENTENNIAL EDITION**

THE LOWELL PRESS, INC. • Kansas City, Missouri

BOOKS BY DONALD R. ORNDUFF

*The Hereford in America*

*Historical Overview
of George F. Ellis'
Bell Ranch as I Knew It*

*Casement of Juniata*

*The Story of Parker Ranch*

*The First 49*

FIRST EDITION



Library of Congress Cataloging in Publication Data

Ornduff, Donald R.
The first 49.

1. Hereford cattle–History. 2. Cattle breeders–United States–Biography.
3. Beef cattle–United States–Breeding–History. I. Hereford Heritage Hall.
II. Title. III. The first forty-nine.
SF199.H4069 1981 636.2'22 81-84498
ISBN 0-913504-73-4 AACR2

*To the Breed Builders—*
*Past, Present and Future*

*Medallions (both sides shown) presented to each honoree (or representative) upon induction into the Honor Gallery of the Hereford Heritage Hall.*

# Contents

# THE CONCEPT

By LEE CAMPBELL
1978 President,
American Hereford Association
(As announced in 1978
at dedication of the Hereford Heritage Hall)

As one reviews the rich and illustrious history of the Hereford breed from its beginning in 1742 when Benjamin Tomkins began identifying individual animals, keeping records and planning matings to produce desired results in animals of beef-making traits, he is impressed with the great number of people who have been involved. Even more impressive is the caliber of those who may be considered as leaders in the various facets of the Hereford industry, people who through their dedication, integrity, sound judgment, determination and often force of personality, coupled with scarcity of mistakes, have guided the Hereford breed to its present position in America.

While every consideration is given to the great animals that stand out as pillars of strength in breed progress, it is the people who have made decisions, made selections, planned matings, exhibited cattle, performed research, perfected marketing programs, improved nutrition and generally guided the industry through the various trends and cycles that have constantly challenged those in agriculture. Thus, it is not only to great animals but also to great people that we are indebted for our breed of today and the stability of our business.

It is, therefore, very appropriate that the directors of the American Hereford Association have endorsed a program to focus permanent recognition on the truly outstanding leaders of the past, present and future for their contribution to the progress of the industry in their time. Each person who is so signally honored will be inducted into the Hereford Heritage Hall housed within the Association's own building. A special plaque for each inductee will be included in the special Gallery of Honor.

So it is with deep gratitude and the heartfelt thanks of all of those who make up the Hereford industry that I, particularly in behalf of the Board of Directors of the American Hereford Association, dedicate the Hereford Heritage Hall. This I do in memory of those of yester-year and the present who have contributed markedly to the breed's accomplishments, and in recognition of the fact that future inductees are, and will be, continuing to make our business the fascinating, challenging and highly important phase that it is of the world's agricultural economy.

# Introduction

By B. C. SNIDOW, Chairman
Hereford Heritage Hall Committee

How fortunate it is that the Hereford industry has in its midst a historian of such an extended background as Don Ornduff, whose talents include the ability to narrate and bring to his readers the wealth of intriguing facts that are a part of the Hereford lore in America.

Starting adult life in the latter 1920's just after college graduation, Don Ornduff began his tenure in the Hereford field by becoming associated in 1930 with *The American Hereford Journal*, a connection that has lasted for over 50 years. During that time he has written countless thousands of words encompassing practically every phase of Hereford activity, the breed's history and its march to preeminence in the western United States.

Before retiring after 10 years as managing editor and 25 years as editor of *The American Hereford Journal*, he wrote *The Hereford in America*, one of the four breed histories that have been published in this country. The first, by T. L. Miller, and entitled *History of Hereford Cattle*, was published in 1902 after Miller's death. The second, *The Story of the Herefords* by Alvin H. Sanders, appeared in 1914, while the next, written by John M. Hazleton, was issued in three editions in the 1920's and '30's. Ornduff's book, *The Hereford in America*, was published first in 1957, with subsequent editions in 1960 and 1969, all of which were sold out and thus are no longer readily available.

When the American Hereford Association's Heritage Hall Committee established the framework for the Hereford Heritage Hall, plans included an Honor Gallery to be composed of a portrait plaque for each person selected for special recognition for contributions to the Hereford industry. Also included in the plan was a historical library to include a biographical sketch covering highlights in the Hereford activities of each honoree and noting the basis for the special distinction accorded.

Who else but Don Ornduff would the committee ask to prepare the story of each honoree? His thorough acquaintance with Hereford history, including personal contacts with many of those involved in it through the years, plus his long-time study of the careers of the cattlemen and Hereford breeders of earlier times, made him easily the first choice of the committee. Fortunate, indeed, it is that Ornduff was willing to undertake the task and fit it into his busy schedule of research and writing.

This is how *The First 49* came about. It is hoped that readers of this volume will enjoy learning about those who are included and why they merit representation in the Hereford Heritage Hall. That the validity of the biographical sketches of the first 49 honorees are of significance in the history of the Hereford breed lies in the fact that history in any field is based largely upon the personal accomplishments of the people engaged in a specific area of activity. While the Hereford breed is made up of cattle, the Hereford industry is essentially a people proposition because people are responsible for its leadership, its trends and continuity.

In this volume we in the Hereford field are indebted to Don Ornduff for these perceptive sketches concerning some of the leaders who have contributed significantly to an on-going Hereford industry for these many, many years.

# Foreword

THE story of almost any institution, or industry, or profession, or breed of cattle is mainly a story of people. When the latter provide intelligent, inspired guidance, in combination with tenacity and industriousness, unusual accomplishment in whatever the chosen field may be is the almost certain result. When these elements are absent, not only is even the *status quo* unlikely to be maintained, but slippage or deterioration is almost certain to ensue. That is nature's way.

Hereford progress through the years has been the product of people possessing ideas of what they *wished* to do and visions of what they believed they *could* do. This book's purpose is to focus attention, as the American Hereford Association marks its Centennial, upon 49 men who, through conscientious and sustained effort became lastingly significant benefactors of the Hereford breed. Not only is it hoped that their stories may yield elements of broad interest relating to the overall history of the breed, but also that others of newer generations may find encouragement in their examples and inspiration in the results of their effort to the extent that they may choose to "go forth and do likewise." If it succeeds in this endeavor it will have served a noble purpose.

In reflecting upon "the first 49," it is significant that a substantial number of them were introduced to Herefords on pastures and in feedlots where the white-faced cattle, frequently crossbreds, usually mingled with cattle of other colors, and often with those of mixed colors. So well did the Whitefaces perform as beef-makers on grass, in comparison with the others, that neighbors of these early owners almost automatically were converted to what was then a new breed to many of them. This gave impetus to the Hereford advance through the cattle country, resulting in an advantage which continues to this day.

Of comparable influence in the breed's beginning years was Hereford performance in farm feedlots. So superior were the white-faced beeves in turning the farm crops of those times into palatable beef that farmers and stockmen who previously had been inclined toward other kinds gave their allegiance readily to Herefords; and became outspoken proponents of the breed, to its ultimate great advantage. And this, of course, spurred the early purebred breeders to intensify their endeavors both in production and distribution.

It has been my privilege through the years to cross paths with two-thirds of those who thus far have been inducted into the Hall's Honor Gallery. With some in my earlier days it was little more than a nodding acquaintance, but one factor which seemed dominant in all of them was that through their activities they displayed a clear sense of purpose and dedication. There is no doubt that this characteristic also was a dominant element in the make-up of the pioneer breeders in the industry's beginning years in America. And that it will find similar application in those who will be inducted into the industry's Honor Gallery year by year in the decades ahead.

Considerable thought has been given to the arrangement of this book's contents. The biographical sketches could have been presented in alphabetical order. On the other hand, a chronological presentation tends to highlight, for those who may read the book from front to back, how a newer breeder's success was established on the basis of an earlier breeder's accomplishments.

The same factor is applicable also to bloodlines, from generation to generation, in a completely logical progression. This may not always be immediately apparent, but such connections are sufficiently frequent to have warranted a decision in favor of presenting these sketches in more or less chronological order, not on a basis of age of the honorees but rather with their eras of activity and influence as the guiding factor. The 49 might well have been divided into three segments: l, The early years; 2, the middle years, and, 3, the later years. But inevitably there would be overlapping, and so it is left to each reader to make his or her own final classification.

For those who do not choose to read the sketches in order from the first to the last, but rather to pick and choose, there is the customary alphabetical index which will provide quick guidance to the names possibly of more immediate interest.

And now, my thanks to all those who have contributed in any manner to the product of this endeavor. Such assistance has been invaluable.

DONALD R. ORNDUFF

October 31, 1981
Kansas City, Missouri

# THE FIRST
# 49
## PERSONALITIES
IN THE HONOR GALLERY OF THE
AHA'S HEREFORD HERITAGE HALL

community, the boy next found work in a store in nearby Richmond. At the age of 14 he became a clerk in the high court of chancery where he served four years, studying law in his spare time. After a year in the office of the attorney general of Virginia he received his license to practice.

In 1797, when he was 20 years old, the young lawyer packed his saddle bags and headed for the new land of Kentucky. From the time he was a small boy he had admired the country homes and farmsteads of some of the well-to-do landowners of Virginia. He had seen that the ownership of land gave the possessor a certain distinction and set him out among his fellows. Young Clay resolved then and there that some day he would have land of his own.

That, perhaps, was one of the reasons why the young lawyer decided to seek his fortune in Kentucky, the "far west" as it was then called. The goal was attractive both as to establishing a law practice and the acquisition of land. So, as he rode into Lexington, then a town of some 1,600 persons, he no doubt had one eye on the rolling countryside, while at the same time he surveyed the group of log cabins in the town's center looking for a possible location for a law office.

Not until some nine years later was he able to make a start on the realization of his land-owning ambition. Almost from the time he had ridden into town from the east he had admired a tract lying a mile-and-a-half to two miles east of Lexington. Finally he was able to purchase acreage there which eventually was to be increased to some 600 acres. He took pride later in saying that it had been acquired entirely through "his own labors."

With his legal career successfully launched, Clay, by then in his early 30's, decided that the time had come to build the country home of which he had

*The house at Ashland as it looked when Henry Clay lived there. The original house was dismantled in the mid-19th century and then rebuilt on the same basic design, continuing to stand as this is written.—Photos courtesy the Lexington-Fayette County Historic Commission.*

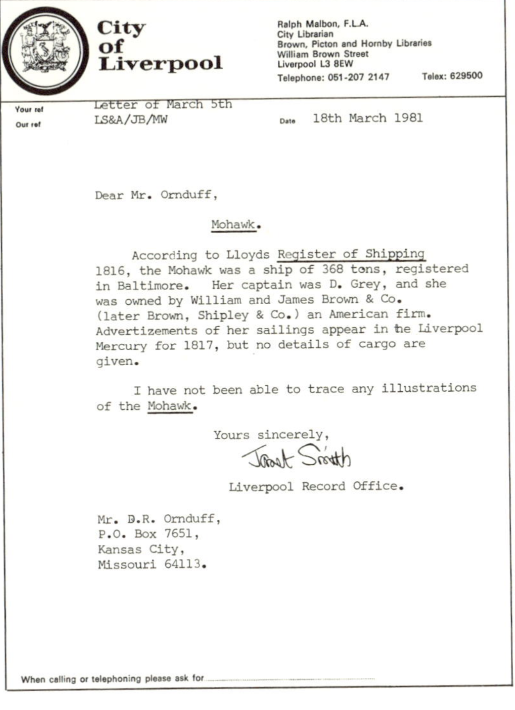

City of Liverpool

Ralph Malbon, F.L.A.
City Librarian
Brown, Picton and Hornby Libraries
William Brown Street
Liverpool L3 8EW
Telephone: 051-207 2147 Telex: 629500

Your ref Letter of March 5th
Our ref LS&A/JB/MW Date 18th March 1981

Dear Mr. Ornduff,

Mohawk.

According to Lloyds Register of Shipping 1816, the Mohawk was a ship of 368 tons, registered in Baltimore. Her captain was D. Grey, and she was owned by William and James Brown & Co. (later Brown, Shipley & Co.) an American firm. Advertizements of her sailings appear in the Liverpool Mercury for 1817, but no details of cargo are given.

I have not been able to trace any illustrations of the Mohawk.

Yours sincerely,
Janet Smith
Liverpool Record Office.

Mr. D.R. Ornduff,
P.O. Box 7651,
Kansas City,
Missouri 64113.

When calling or telephoning please ask for

*An exhaustive search on both sides of the Atlantic indicates that no likeness or sketch of the ship* Mohawk *exists. This letter from the public library of the city of Liverpool, England, from which this ship carried Henry Clay's Hereford purchases to America, shows that it was a vessel of quite limited tonnage.*

dreamed. An imposing brick house arose on the farm which he had named Ashland, and here Henry Clay lived from the time of its completion about 1810 or 1811 until his death in 1852. The original mansion was damaged by fire and was razed some 10 years later, but another Ashland mansion was erected on the ruins of the old, a replica of the original, and stands today surrounded by great old trees, vine covered and in its artistically landscaped setting. This is the shrine preserved by the Henry Clay Memorial Foundation which visitors see today, though the acreage on which the present Ashland stands has shrunk to 20 acres, surrounded by a city residential district.

Henry Clay did not buy his 600 acres merely to become a "front porch" farmer. With an inclination toward both the law and the field of agriculture, Clay decided to follow both pursuits. Just as he refused to be a run-of-the-mine lawyer, he made a deep and continuing study of agricultural matters and kept meticulous accounts of all details of his farming operations. He planned with care and gave to his farm his personal attention except when public duties made his absence necessary. Even then, letters of instruction and memoranda for the guidance of his farm overseers, still preserved, leave no room to doubt that, at home or not, Farmer Clay ran his Ashland Farm.

It was the livestock on his property that gave Clay his greatest pleasure and in which he took the highest pride. Horses, mules, cattle, hogs and sheep roamed his acres, with more than one breed represented in some instances. When he learned that a Kentucky neighbor, widely known as a farmer-stockman, Col. Lewis Sanders, had placed an order in England for a shipment of Shorthorn and Longhorn cattle, he inquired as to the possibility of including in the shipment a few Herefords from the west of England. He had read of this comparatively new breed, and of its efficiency in beef production. After correspondence with an acquaintance, Peter Irving of Liverpool, a brother of the well-known author, Washington Irving, Clay was able to secure two Hereford bulls and two females, which were shipped from Liverpool on the steamship Mohawk with the Sanders cattle to Baltimore. One of the bulls died either on board ship or enroute from Baltimore to Kentucky, but the three which arrived comprised the first Hereford-breeding unit in America.

Careful farm operator that he was, Clay bought modestly, obtaining good quality but at most reasonable prices. Under his original, carefully laid plans, he had intended that those two bulls and two females were to be the nucleus of a test herd that he had in mind. The unfortunate death of one of the bulls made a change in plan necessary. The distance and expense involved in another shipment from abroad precluded this option, so Clay proceeded as best he could with the three Whitefaces which arrived at Ashland.

For a number of years he kept his Herefords separate from other breeds on the farm, but eventually, fearing the effects of continuous inbreeding, he mixed them with his Shorthorns and their identity was lost. With characteristic prepotency they imprinted their trademark of the white face upon hundreds of cattle in the Bluegrass State. Cattle descended from these Clay Herefords became widely known as the "Seventeens," in token of the year of their importation, 1817. And Clay's farmer-neighbors were favorably impressed by the Hereford influence. The winning steers at the Chicago Fat Stock Show in 1878 and 1879 were out of cows of the Seventeens stock, and were said even at that late date to still show the Hereford markings.

Concerning his decision between the Herefords and the Shorthorns, or Durhams as the latter then were sometimes known, Clay later wrote:

"The choice between the two races should be regulated somewhat by circumstances. My opinion is that Herefords make better work cattle, are hardier, and will, upon being fattened, take themselves to market better than their rivals. They are also fair milkers. . . .

"If one has rich, long and luxuriant grass, affording a good bite, and has not too far to drive to market, he had better breed the Durhams, otherwise Herefords."

Thus, from the very beginning of their history in America, the Herefords' easy-keeping, characteristic hardihood and ability to take care of themselves were recognized by Ashland's owner. Indeed, those same traits would come in succeeding generations to be recognized and appreciated across the breadth of America. But Ashland was where it started.

WILLIAM H. SOTHAM

T.F.B. SOTHAM

# WILLIAM H. SOTHAM

ALBANY and BLACK ROCK,
NEW YORK
1801-1884

# T. F. B. SOTHAM

CHILLICOTHE, MISSOURI
1863-1926

THE first importation of Herefords to the United States that resulted in the establishment of a breeding herd on a substantial basis was made by William Henry Sotham in 1840. He was born in 1801 at "The Woodleys," his father's 200-acre livestock farm at Wooten, in the English county of Oxford. After completing his formal schooling as a 14-year-old boy, he took his place as a "hand" on the property, giving much attention to the care of the livestock and eventually to the marketing of it. And it was his experience at the markets that was to be strongly reflected in his later cattle activities.

His early home in Oxfordshire was not far from Herefordshire, the county in the west of England where the Hereford breed originated. The young Sotham thus became well acquainted with the livestock of that locality and noted the outstanding beef-making characteristics of the white-faced cattle he saw on various farms and in the shows and sales of beef animals at the city of

*This was "The Woodleys," at Wooton, Oxfordshire, England, family home of the Sothams and birthplace of William H. Sotham, the earliest importer of a sufficient number of purebred Herefords to comprise a significant breeding herd in the U.S.A.—Sotham photo.*

Hereford. Likewise, he was greatly impressed by the Herefords he later saw at the great Smithfield beef cattle show in London.

In 1832 Sotham left his native land for America and almost immediately became identified with livestock interests on this side of the ocean. First, he was employed on a farm at Euclid, Ohio, to manage the place and to handle the production and sale of the livestock on it. After two years he returned to England for a visit, and on this occasion and during subsequent trips home he continued his visits to the stock farms of the area, the livestock shows and the markets. And always when he returned to America he lost no opportunity to extol the merits of the Herefordshire cattle.

When in due course he became a cattle buyer for Ebenezer Wilson, a New York meat packer, he had new and extended opportunities to see and study American beef cattle and to compare them with the bullocks he had observed in his homeland. The American cattle suffered by comparison, and in his customary outspoken manner the Englishman did not hesitate to say so, even though his frankness did not endear him to proponents of the then dominant Durhams. He did, however, succeed in arousing the interest of his employer, the upshot of this being that Wilson sent him to England to select an importation of the Hereford breed of which Sotham had spoken so convincingly.

Thus it was that William Henry Sotham in 1840 brought to America 22 purebred Herefords, mostly representing the breeding of the herd of John Hewer, one of the most noted cattlemen of his era. This lot consisted of 12 cows, plus some heifers and calves, the caliber of the group being indicated by the fact that one of the cows had been champion at the Royal Agricultural Society's show in England the previous year.

A problem, however, had arisen in Sotham's absence. Wilson had suffered financial reverses to the extent that he was unable to pay for the shipment, but he had succeeded in arranging for Erastus Corning of Albany, New York, to assume this responsibility. The Albany business and civic leader, who already had a herd of Shorthorns, thus became the principal owner of the shipment, with Sotham having a small interest on the basis of the down payment he had made before the cattle left England. This partnership continued for seven years, when Sotham bought the bulk of the herd and moved it to a farm he had acquired at Black Rock, in the vicinity of Buffalo, New York.

Thus, William Henry Sotham became a unique benefactor of the Hereford breed in America, continuing his activity until past 80 years of age. He knew most of the active breeders of his time, and contributed to the successes of many of his contemporaries. He had courage and the militant ability to make his vision for Herefords become a reality. Of him, Alvin H. Sanders, editor of *The Breeder's Gazette*, leading livestock weekly of its time, said:

"Sotham was a thorough-going believer in the superiority of the Hereford–a courageous pathfinder. He was as indefatigable in defending Hereford interests as he was active in assailing those who failed to confess what he held to be the true faith; and while he was for many years a sort of John the Baptist, crying vainly in a forbidding wilderness, his voice was heard by some who harkened and subsequently became pillars in the Hereford church."

He was zealous in any cause he espoused, and it was said of him that he was belligerent to a fault. Nor was he a practitioner of the gentler arts of diplomacy. He had an indomitable will, strong convictions, rare courage and a love of the practical.

"There is no science, no study so pleasing, so substantially gratifying to the mind of man," he once wrote, "as to fully develop the good points of his animals through his own superior skill and management." He went on: "He who trusts entirely the opinion of others will never make a breeder. Results from his own experience must be his guide, and when a breeder arrives at the highest point of excellence, his name spread far and wide, it is a very difficult matter to keep it there. Prosperity is apt to make men careless, which is almost certain to create degeneration, and when this takes place the downward steps are long and rapid."

Sotham was a man ahead of his times, but that does not fully describe him. He was a man of boundless energy and unusual ability and possessed the vision to see what Herefords could do in putting heavier quarters and a thicker covering of flesh on America's beef cattle. He consistently urged organized unity, as indicated by this statement, which he repeated in essence again and again:

"The breeders of stock and tillers of soil ought to be intimately united or the anticipated improvement in husbandry will end in disappointment."

He cited the advantages which he believed were certain to accrue from the formation of appropriate associations and societies, and it must have been highly gratifying to him that one of the early societies, as urged by him, was the American Hereford Cattle Breeders' Association (now the American Hereford Association), formed in 1881, when he was 80 years old. Recognition of the worth of his work over so long a period, and of his genuine accomplishments in behalf of the breed, was given at the association's first meeting in June, 1881, when he was elected a vice-president and was presented with a copy of Volume I of the American Hereford Record by the association's president, C. M. Culbertson, who hailed him as "the old champion and original Hereford advocate." This volume contained a full-page picture of Mr. Sotham as a special feature.

Tom Sotham, more formally known as Thomas Frederick Beaubois Sotham, succeeded his father and inherited many of the senior Sotham's personal characteristics, notably his fiery and aggressive spirit in his advocacy of Herefords. Not only did he absorb well the lessons learned at his father's knee, but as a mere youth he had the added advantage of close contacts when employed by three distinguished early-day Hereford leaders. At the age of 15 he was sent to T. L. Miller's Highland Stock Farm at Beecher, Illinois, to learn the practical part of a herdsman's duty; at 16 he became herdsman for Tom Clark at Evergreen Stock Farm at Beecher, and at the age of 17 he was named

*This medal (both sides shown) was presented by the New York State Agricultural Society to William H. Sotham in recognition of his accomplishments as a beef cattle improver through the Hereford breed during the middle years of the 19th century.*

*T. F. B. Sotham was a great practitioner of publicity and promotion. This drawing by the animal artist, Cecil Palmer, was widely used by Sotham in advertising his herd to the far corners of America, and beyond. At left is the $5,000 Sir Bredwell, Omaha Trans-Mississippi and International Exposition champion, which was sold to Col. C. C. Slaughter of Texas. Above is Thickset, many times a winner at leading shows between 1897 and 1899, which was purchased for $5,100 by William Humphrey of Ashland, Nebraska. Both were sired by Corrector, at right, the noted Sotham herd bull which was valued by his owner at $100,000.*

head herdsman for Tom C. Ponting at Moweaqua, Illinois, keeping up the herd records at night after finishing the day's work with the cattle and in the fields.

He first engaged in the Hereford business on his own personal behalf at Flint, Michigan, and a little later became a partner in the firm of Sotham & Stickneys at Pontiac, Michigan. There, in 1883, he got out the first Hereford catalogue issued with footnotes. When that partnership was dissolved in 1890 he put out an elaborate, extensively illustrated catalogue which probably was the second ever published for a beef cattle sale containing five-generation pedigrees. In general use then were the so-called long-form tabulations with the main emphasis on the maternal line of descent. These innovations were manifestations of an exploring and independent mind.

Tom Sotham shortly after that Michigan dissolution sale purchased the prominent herd of J. O. Curry of Aurora, Illinois. In 1893 he moved his cattle to Chillicothe, Missouri, where his Weavergrace Farm herd during the

*T. F. B. Sotham here was delivering Sir Bredwell to Kansas City, the first step in the bull's journey to Col. C. C. Slaughter's ranches in West Texas. According to reports at the time, the trip involved "800 miles by rail and 80 miles on foot."—Sotham photo.*

ensuing 10 years became one of the most prominent and successful of its time. His show herds attained remarkable success, the climax perhaps being reached at the Trans-Mississippi and International Exposition at Omaha, Nebraska, in 1898 when his Sir Bredwell was acclaimed Sweepstakes Bull of any age.

The following March, Sotham sold 46 head at auction for an average price of $516, the highest that had been recorded on a Hereford auction in the United States in 16 years. Sir Bredwell topped the event in going to Col. C. C. Slaughter, Dallas, Texas, for his West Texas ranches at $5,000, setting a new record for an American-bred Hereford bull at auction. In 1900 Sotham was elected president of the American Hereford Cattle Breeders' Association for the ensuing year.

As a firm exponent of the efficacy of the showring as a herd and breed promotion medium, he continued to exhibit at leading events during the early 1900's. Seeking to expose the merits of the breed to stockmen of the southeastern states, he took his herd to an exposition at Charleston, South Carolina, where his cattle contracted tick fever. Several of them, including a champion show and herd bull, died. This and the loss of two other stock bulls resulted in financial difficulties, which caused the herd to be closed out in December, 1903.

Sotham later engaged on a limited scale in breeding Herefords in Michigan. Still later, in association with his three sons, William, Harold and

Thomas, he conducted a successful sale management service, his firm handling among other notable auctions of the period the dispersion sale of Gudgell & Simpson at Independence, Missouri, in 1916. The last survivor of the sons, Thomas, died in Kansas City in the fall of 1978 at the age of 89, only two weeks before he was to have represented his father and grandfather at their induction into the Honor Gallery of the American Hereford Association's Hereford Heritage Hall.

T. F. B. Sotham's main contribution, however, in the light of history, was not so much in the influence his productions had upon subsequent generations of Herefords as to the advances that may be attributed to his vigorous promotion of the breed and the wide dissemination of Herefords, both registered and commercial, to which his activities contributed. He was an all-out exponent of Herefords wherever he went.

A contemporary of Tom Sotham, and a close observer of his activities, was Frank S. Hastings, who worked for Kirkland B. Armour in Kansas City and then managed the SMS Ranches in West Texas. In his book, *A Ranchman's Recollections*, Hastings paid this tribute to Sotham's foresight:

"For many years the agricultural press has at times, in a casual way, credited me with being a pioneer in pushing range-bred calves to be matured as finished beeves in the corn belt at an age not exceeding 20 months. While I have devoted 20 years of my life to that work, the idea was in the main obtained from T. F. B. Sotham long before I had any thought of becoming identified with range work, and, while many feeders undoubtedly were trying

*Few herds were more prominent and successful at the turn of the century than the Weavergrace Herefords of T. F. B. Sotham. A portion of the cow herd was pictured here at the farm headquarters northeast of Chillicothe, Missouri—Sotham photo.*

it in a small way, even before Mr. Sotham took it up, he was the first to get behind it in any definite way. While it was sure eventually to come, his initial work brought its first impetus, introducing it as a distinct phase of the feeding industry years before its natural evolution could have brought it about."

Hastings also acknowledged Sotham's effective activity in sending Hereford bulls to the range country in these words:

"When I began my work with the Armour herd, Tom Sotham was a 'live wire' in extending the use of registered Hereford bulls into the range. He made many visits to range herds, came in contact with the big and little men of the range, recognized the rapid improvement that had been made, initiated some experiments as to the outcome, and became convinced that the market for registered bulls would increase directly in proportion to the benefits that ranchers received from their use. My attention was first called to his work when he made an effort to induce a number of ranchmen to contribute to a public sale 100 good steer calves, or, say, two carloads each, to be placed by him [Sotham] and developed into 'baby beef.' "

It was Tom Sotham's conviction that such an effort would demonstrate to corn belt cattle feeders the value of improved breeding in the cattle they fed out, and that the importance of the continuing relationship between the purebred breeders, the commercial ranchers in the range country and the cattle feeders who were in the business of finishing beef for market thus would become widely manifested. There is no doubt that T. F. B. Sotham was an "idea" man, and if he ran into early disappointments, he continued to try again and again until his point was made.

A distinctly worthwhile contribution to the literature of the breed was a compact Hereford herd-promotional item and history, *A Treatise and Handbook on Hereford Cattle*, which he published and circulated widely during the peak of his Weavergrace operations. A unique feature of this 408-page volume was Sotham's Star List or Merit Record which he compiled, and which contained not only miniature illustrations of most of the famous Herefords of earlier times but also listed their winnings in the principal shows of England and the United States. The book also contained the pedigrees and produce lists of the Weavergrace breeding herd.

Sotham also assumed the responsibility of publishing the first American Hereford history, T. L. Miller's *History of Hereford Cattle*, which that pioneer Hereford advocate had written but never published. This book was brought out by Sotham in 1902.

# T. L. MILLER

BEECHER, ILLINOIS
1817-1900

T. L. MILLER of Beecher, Illinois, played a major role in guiding the Hereford breed into its first significant era of expansion, beginning around 1870. His determination, energy and resourceful leadership, which had been demonstrated earlier in successful business enterprises, proved to be a vital factor in bringing the breed to the wide and favorable attention of American stockgrowers.

Born at Middletown, Connecticut, in 1817, Miller went to Ohio in 1842 where he was in business until 1856, when he moved to Chicago. There he engaged successfully in the fire and life insurance business, meanwhile acquiring farm land in Will County, Illinois, some 40 miles south of the city. He commenced to improve this property with buildings in 1862, continuing at the same time to add to his holdings until they totalled nearly 1,100 acres. When a railroad was built through the area he laid out the village of Beecher on a portion of his property.

In March, 1870, he closed out his business in Chicago and went to live at Beecher on what he had named the Highland Stock Farm. The names of the farm and the town became widely familiar in agricultural circles during the ensuing decade, mainly because of the aggressive campaign he waged from there in behalf of Herefords.

*Success 2 was the herd bull which brought acclaim to the early-day Hereford-breeding effort of T. L. Miller. This artist's pen-and-ink sketch obviously is exaggerated, but there is no doubt of the bull's influence in contributing to the early acceptance of the then new breed in the Central states and elsewhere.*

Meanwhile, a young man from Herefordshire, William Powell, who was a descendant on his mother's side from the Tomkins family which is generally credited as having been the founders of the Hereford breed, and who had gained a considerable reputation in Ohio as a stockman, came to Miller's favorable attention. The result was that Miller negotiated with Powell to become manager of the Highland Stock Farm where the herd actually was begun under the ownership name of Miller & Powell.

William Powell later became a pioneer in the introduction of Herefords into Texas, first associating himself with B. C. Rhome of Fort Worth, and later operating a purebred establishment of his own at Childress for three years before eventually establishing his Hereford Home herd on a large ranch at Channing, Texas.

But back to Illinois: Miller & Powell's first investment in Herefords consisted of a one-half interest in three cows and two bulls purchased in Ohio in 1872. Later in the same year came others, including the five-year-old Sir Charles, already proved to be a superior show bull and sire, acquired from F. W. Stone of Guelph, Ontario. The latter cost the sum of $1,000 in gold, which equalled $1,300 in the currency of the times, making Sir Charles the first high-priced Hereford brought into what then was generally considered "the west." After five years of service in the Highland herd the old bull went to the butcher still weighing 2,550 pounds, bringing a price per cwt. equal to that paid for the best bullock sold on that day and dressing 70 per cent.

Not satisfied with the best that could be bought from existing American and Canadian herds, Miller within a year or two after his start, imported 170 head from leading English herds representing the more prominent bloodlines of the times. Included in this shipment was Success 2, linebred to the great Sir David 68. Success 2, as a yearling, became a first-prize winner at four leading fairs in the central states in 1874.

Miller had learned to appreciate the value of advertising while in the insurance business, and made liberal use of such mediums of publicity as were available. Practically all of the agricultural papers of the day favored an older-established breed, the Shorthorns, but this did not deter Miller from using these publications in his aggressive campaign to promote the interests of the newer breed.

Eventually the pressure upon the agricultural and livestock press from the opponents of Herefords became so great that Miller decided the only way to secure a fair hearing for the Whitefaces was to establish a new magazine. Thus, *The Breeders' Journal*, which he founded in 1880 and published monthly until 1887, came into being, and proved to be an immensely powerful factor in advancing the interests of the breed and bringing its merits to the attention of ranchmen located in states farther west.

As already indicated, Miller also recognized earlier than most of his contemporaries the very distinct value of a good show herd, and was represented as early as 1872 by entries at the Illinois State Fair and the St. Louis Fair. He had no complaint regarding the breed classes, but when it came to interbreed competition for sweepstakes awards he felt that he was not getting justice from the fair boards which were dominated by Shorthorn breeders. But he proved himself a good sportsman and continued to exhibit his Herefords. Success crowned his effort when in 1876 his herd won the sweepstakes award over all breeds in the show at Cleveland, Ohio.

This same year, at the Centennial Exhibition at Philadelphia, Success 2 was grand champion bull and Miller's herd, under William Powell's care, was awarded a medal as the first-prize Hereford herd exhibited. Of the Miller herd entry, the judge, Thomas Duckham of Hereford, England, who edited the English Hereford Herd Book for 21 years during its formative period, said: "The animals comprised in this herd display fine character, form and quality, and are entitled to rank as very first-class specimens of the Hereford breed."

It is said to have cost Miller $2,000 to take his show herd to Philadelphia, and as no cash prizes were awarded the entire amount had to be charged up to advertising. If this venture proved to be an excellent advertisement for Miller's herd, it was an even better one from the standpoint of the breed. It brought Herefords favorably to the attention of stockmen in attendance at the Philadelphia event who previously knew of them only by hearsay. Among

INTERNATIONAL EXHIBITION, PHILADELPHIA, 1876.

AWARD TO THE "HIGHLAND" HERD.

The United States Centennial Commission has examined the report of the Judges, and accepted the following reasons, and decreed an award in conformity therewith:

PHILADELPHIA, Jan'y 9, 1877.

REPORT ON AWARDS.

Product---Hereford Cattle. Name and address of Exhibitor---T. L. Miller, Beecher, Ill.

The undersigned, having examined the product herein described, respectfully recommends the same to the United States Centennial Commission for award, for the following reasons, viz.:

*First*---Hereford Herd. The animals comprised in this herd display fine character, form and quality, and are entitled to rank as very first-class specimens of the Hereford breed.

*Second*---Hereford Cow Katie, Hereford Heifer Charlotte, Hereford Heifer Victoria, Hereford Bull Seward. For good character, form and quality, we consider them superior specimens of the Hereford breed.

*Third*---Five Hereford Heifers: Prairie Flower, Faith, Dolly Varden 2d, Mary Hughes, and Abbie; Hereford Cow Grace, Hereford Cow Dolly Varden, Hereford Bull Success. Their exceedingly fine character, form and quality entitle us to consider them to be first-class specimens of the Hereford breed, and worthy of our highest commendation.

*Fourth*---Hereford Bull Royal Briton. Commended for his good character and quality of flesh, combined with great scale.

T. DUCKHAM,
Signature of the Judge.

APPROVAL OF GROUP JUDGES.

ASHBEL SMITH,
T. C. IRISH,
H. C. MEREDITH,
COLVIN CAMERON,
M. WILKINS.

A true copy of the record.

FRANCIS A. WALKER,
Chief of the Bureau of Awards.

Given by authority of the United States Centennial Commission.

A. T. GOSHORN,
Director-General.

J. L. CAMPBELL,
Secretary.

J. R. HAWLEY,
President.

*T. L. Miller's display at the Philadelphia World's Fair in 1876 brought Herefords favorably to the attention of many visitors there for the first time. Among those who viewed the Miller Herefords at the fair and decided almost immediately to enter the ranks were the Gudgell brothers, Charles and James, and their partner, Thomas A. Simpson. Another was Capt. W. S. Ikard of Texas, one of the first to take Herefords to the Lone Star state. Thus evident is the influence of the Miller herd, which was appropriately acclaimed by officials of the Exhibition, as shown above.—Reproduced from Miller catalogue in the author's personal file.*

*This unique stock barn at T. L. Miller's Highland Farm at Beecher was an area landmark.*

these were numerous western and southwestern range cattlemen who were impressed by the possibility of increasing their profit by improving the quality of their output. The sight of the sturdy, smooth-bodied, thick-fleshed Herefords at the Philadelphia exhibition gave many a western and southwestern ranchman a new vision of the range cattle herds of the future.

One of these visitors was Captain W. S. Ikard of Henrietta, Texas, who after looking over some top-rank representatives of the breed for the first time decided then and there to buy some of them. This resolution led to the establishment of the pioneer purebred herd of W. S. & J. B. Ikard, considered to have been the state's first. The Ikard name continued to figure in Texas Hereford affairs for almost a half-century.

But the sight of the T. L. Miller Herefords impressed three other Philadelphia World's Fair visitors to the extent that their operations during succeeding years changed the course of Hereford history in America for all time. Those visitors were Charles and James Gudgell, brothers, and their partner, Thomas A. Simpson, from Missouri. They were virtually overwhelmed by the excellence of the exhibit, and deemed the Hereford potential in the ranch country, which spread westward from their location near Kansas City, to be so great that they decided forthwith to engage in the Hereford business. Knowing that some of the Miller foundation stock had come from the F. W. Stone herd in Ontario, Charles Gudgell decided to visit the Stone farm on his way home. He bought there a bull and seven females. So favorable was the impression created by these first Gudgell & Simpson Herefords that a year later Charles Gudgell returned to Canada to buy another bull and several more cows and heifers.

Thus, T. L. Miller deserves a large share of the credit for bringing the Gudgells and their partner, Simpson, into the Hereford orbit. Without his initiative and enterprise, it might never have happened. And the Hereford breed's history in America might well have been far different.

As a student of Hereford bloodlines and inheritance, and with an awareness of the importance of the influence of both sides of a pedigree, Miller deplored the use of pedigrees which unduly emphasized the maternal line of descent–the so-called long form which carried the bottom line back to the eighth or tenth generation, to the exclusion of the other lines. He therefore pioneered in the use of the fully tabulated five-generation pedigree, and the catalogue of his sale on July 22, 1885 is believed to have been the first in which such pedigrees were presented. A copy of this catalogue is in the author's file.

Even if Miller had not attained distinction through his Hereford breeding and promotional activities, he would have deserved a rank of honor in the industry through having established the American Hereford Herd Book in 1880, with Volumes 1 and 2 having been issued privately under his direction in the name of his Breeders' Live-Stock Association. Since these records were vital when the Hereford breeders formed an official association in 1881, Miller shortly made a proposition to sell the Herd Book to the American Hereford Cattle Breeders' Association, the transaction being consummated in the fall of 1882. The purchase price was $5,000.

Also, to Miller's credit is the fact that he was the author of the first Hereford history published in America. Entitled *History of Hereford Cattle*, it had been written during his latter years, but was unpublished at the time of his death. T. F. B. Sotham believed so strongly that this work was an asset to the industry and would provide inspiration to the breeders for years to come that he personally arranged for its publication and initial distribution.

It also deserves to be noted that T. L. Miller's son, T. E. Miller, was named secretary of the Hereford Association at its organizational meeting in 1881, and occupied that position until 1884.

All of these things contributed to the stature of Timothy Lathrop Miller as a leading pioneer in the development and progress of the Hereford breed in America.

# C. M. Culbertson

CHICAGO and NEWMAN, ILLINOIS
1819-1901

CHARLES M. CULBERTSON was considered the most powerful recruit attracted to the Hereford ranks in the Central states through the influence of T. L. Miller, the latter one of the earliest and most vocal exponents of the Hereford breed of his time. A retired meat packer and a man of first importance on the Chicago Board of Trade as well as one of the original incorporators of the Chicago Union Stock Yards, Culbertson had a tract of 2,340 acres of valuable land at Newman, Illinois, and when he decided to stock it with Herefords a convert had been made whose example was bound to bring powerful support from others.

Of Scottish descent, Culbertson was born in 1819 in southeastern Indiana where educational opportunities were so limited that his formal schooling ended when he was 14 years old. Relying thereafter upon his own perceptions and industry for the elements of knowledge which he acquired, he left home at that age and went to Newport, Indiana, where he clerked in a store for the next eight years. His beginning salary was $80 a year, and eventually rose to $200 a year.

But his dynamism in this job brought him to the favorable attention of David A. Jones with whom he entered into a partnership to carry on a general merchandising and pork-packing enterprise at Newport, in west-central

Indiana, which was continued there until 1856, and thereafter at Chicago until 1864 when the Jones interest was sold to Moses Fowler and Adams Earl of Indiana and Lyman Blair of Chicago. The Fowler and Earl interests subsequently were sold to the other principals, and the firm of Culbertson, Blair & Co., by the early 1870's, was rated a multi-million dollar enterprise. Culbertson himself gained wide recognition as an innovator in the meat-packing business. And it is regarded as highly probable that his packinghouse observations contributed to his decision to engage in the purebred Hereford field.

Culbertson's first purchase was made in 1877 from Miller and consisted of 15 females. His farm, soon to become widely known as Hereford Park, shortly established its reputation as the premier breeding concern of its kind in Middle America. The owner had been feeding steers for the slaughter market for a number of years, and had accumulated a good herd of cows, mainly of Shorthorn extraction. They represented mainly the bloodlines assembled by John D. Gillett of Elkhart, Illinois, acknowledged early-day cattle king of the Cornbelt.

These cows provided excellent material for matings with Whiteface bulls at the Culbertson farm, and out of such a cross came Culbertson's famous Chicago Fat Stock Show champion steer, Roan Boy. That bullock convinced many a farmer and stockman that the get of the Hereford bulls produced a highly superior product for feedlot purposes, and Whiteface popularity thereafter advanced by leaps and bounds, not only in the Cornbelt but far beyond.

As Culbertson's appreciation of the Hereford breed's potential deepened, so did his determination to build a herd of significant stature. He had come to appreciate the skill and judgment of the Herefordshire-born George F. Morgan who was employed as herdsman by T. L. Miller, and in 1879 he persuaded Miller to let Morgan go to England to select for the Hereford Park farm a choice herd-bull prospect and some heifers. The venture proved historic.

Morgan, delighted to revisit his homeland, located and bought for Culbertson the bull Anxiety 2238, then a little more than two years old, from his breeder, T. J. Carwardine of Stocktonbury farm, in Herefordshire. Anxiety in 1877 had won first at the English Royal and was a prize winner at numerous other shows, including the following year's Royal. The price paid for the bull was the English equivalent of approximately $1,000, a long figure for the times, on top of which were the expenses involved in shipping him and the other Herefordshire acquisitions to the Culbertson farm.

It has been related that on the day these cattle were to be loaded aboard

*In the days before photography, artists' sketches depicted their subjects with pen and brush. In the overstated style of the times, this is how an artist saw Anxiety, the founder of a famous Hereford family, at the age of three years on the Culbertson farm.—Dewey sketch.*

ship at Liverpool, England, Morgan met by chance on the streets there an old friend, John Gosling, later to become an important figure in the cattle trade of the American Middle West. Morgan said to him, according to accounts passed down, "John, come with me and I will show you the best bull you ever saw." But Morgan could not possibly have had the clairvoyance to foresee the bull's potential greatness, nor obviously, did Carwardine, else he would not have let the bull go.

In Culbertson's show herd during the latter part of 1879 and in 1880, Anxiety proved to be the sensation of the hour. No such Hereford bull up to that time had been seen in America. Breeders in the United States were accustomed to the rangier, rougher sort. It was but natural, therefore, when Anxiety came along in the Culbertson show string for all to see, displaying his great thickness of flesh smoothly laid on, his rare finish and his great quarters that he should have made a tremendous impression upon a generation of cattlemen who watched as he won top breed award at six successive important fairs. Breeders who saw him were unaware up to that time that such symmetry and balance of parts were to be found in the Hereford breed.

Anxiety's untimely death as the result of an intestinal blockage at the end of the 1880 show season, after making such a deep impression upon the industry, was an incalculable loss to the breed, all the more so because so few calves were born to his service in America. Had his owner anticipated this situation he probably would have kept him busily in service rather than in the show string on the road. In any event, the demand for sons of Anxiety far exceeded the supply.

Fortunately, however, Anxiety had been used in the Carwardine herd to a limited extent before his shipment to America, and his greatest fame flows, as a matter of fact, from the subsequent importation and use in the U.S.A. of two of his sons that had not been born at Carwardine's farm at the time of Morgan's visit. One of these two was Anxiety 4th 9904, that was destined to revolutionize the breed through his service in the Missouri herd of Gudgell & Simpson. One hundred years after Anxiety's importation, at the time of this writing, virtually all of the Herefords in the United States are descended, in the main, from Anxiety 4th by Anxiety.

Had Culbertson's only claim to fame rested upon nothing more than the fact that he imported Anxiety, and thus brought this Hereford family to the attention of the burgeoning American Hereford industry, this would have been quite enough. For it was the impression Anxiety created in this country that resulted in the importation two years later of Anxiety 4th 9904, to the lasting benefit not only of the Hereford breed but the entire American beef cattle industry.

HEREFORD PARK

CHOCE HEREFORDS AT AUCTION!

THE PROPERTY OF

C. M. CULBERTSON

CHICAGO, ILLINOIS.

On November 17th, during the Fat Stock Show, I will offer at Dexter Park Union Stock Yards, Chicago, a draft of 31 good ones from my herd. 23 Cows and Heifers, three of them sired by

**THE GROVE 3d 2490,**

and 9 in calf to him. 8 bulls from 8 to 20 months old, 4 of them by "The Grove 3d," and all good ones. My pedigrees show the rarest combination of the best strains of blood including that of

*THE GROVE 3d 2490, LORD WILTON 4057, ANXIETY 2238, SIR RICHARD 2d 970a.*

My line of breeding and show bulls headed by The Grove 3d, will be on exhibition at Dexter Park one week prior to the sale. Sale to commence at 1 o'clock sharp. Send for Catalogue to

*C. M. Culbertson, Newman, Ill.*

Col. Judy, and C. H. Capern, Auctioneers.

*In the low-key manner of the times, this advertisement of C. M. Culbertson's 1885 Hereford sale appeared in the November issue of* The Breeders' Journal, *published at Beecher, Illinois. With a great reputation already established in England when Culbertson imported him, it is obvious that The Grove 3d was a bull of wider renown than any other then featured.*

This, however, was by no means Culbertson's only contribution to the building of the breed. He also brought across the Atlantic two sons of Lord Wilton, a sire which achieved fame in the Carwardine herd. One of these was Bowdoin, a noted show bull which became the sire of a Culbertson steer which in 1890 captured the championship of the prestigious Chicago Fat Stock Show. In 1883 at $4,150, a new record for Herefordshire at the time, came The Grove 3d, hailed as the "mighty" son of Horace. Subsequently in the Culbertson herd came the mingling of the blood of these three great sires to further improve and solidify the position of the Hereford breed in American beef-making circles.

Culbertson was influential also in other directions. One of these was in successfully bringing Herefords to the attention of well-to-do Chicago businessmen looking for investments in agriculture and cattle production. Several such men engaged in Hereford breeding as a result of his initiative, some of them continuing over a period of years. They became customers of already-established breeders, and, in turn, their herds were sources of supply for others who became interested in Hereford production. Thus, the Hereford industry of that era steadily expanded.

In still another respect Charles M. Culbertson left an indelible imprint upon Hereford history. It was this foresighted Illinoisan who called together the small group of breeders who assembled on June 22, 1881, at the Grand Pacific Hotel in Chicago to organize a breed association with a twofold purpose: first, to maintain records to assure the purity of the breed, and second, to promote the interests of the Hereford industry. The convictions of these men in behalf of their breed, and their arguments in its behalf were crystallized into action when they associated themselves together in that memorable session which marked the beginning of a new–and continuing–chapter in Hereford history.

To Culbertson thus goes a large share of the credit for bringing America's Hereford producers into the close-knit organization known today around the world as the American Hereford Association. It was fitting that Culbertson should have been named as the Association's first president, and that he should have been re-elected to that position in the organization's first annual meeting held November 11, 1881, in accordance with the plan adopted in June. So ably did he guide the infant group that his fellow breeders re-elected him in 1882 and yet again in 1883. No other breeder to this day, a century later, has been elected to the Association's presidency for three successive terms. Quite a tribute to a dynamic Hereford pioneer, whose activity in the field continued until only a few years before he died in 1901 at Hot Springs, Arkansas.

# GEORGE F. MORGAN

NEWMAN, ILLINOIS, and
CHEYENNE, WYOMING
1846-1903

THERE is no doubt that George F. Morgan's activity during the first great expansion era for Herefords in the United States was a major factor in the course of the breed's history throughout the 20th century. Hereford breeders of America will be forever in his debt for having selected in England the bull, Anxiety, for importation, because this acquisition in all probability led to the subsequent importation of Anxiety's son, Anxiety 4th. The significance of this observation rests upon the fact that almost all of the registered Herefords in the United States as of the last quarter of the 20th century descended from Anxiety through Anxiety 4th.

Morgan was born in Herefordshire and, as the son of a tenant farmer, probably was aware of the qualities of this new breed which originated in his native county. But there is nothing to indicate that his father was involved in any way with the Herefords. As with many an English lad of his time, young Morgan decided early to emigrate to the New World, and when he arrived in Cleveland, Ohio, in 1866 at the age of 20 his resources came to a total of only $7.50. But he was ambitious and industrious, which perhaps are more valuable assets than money.

As soon as he could manage to do so, he rented a farm at nearby Elyria,

Ohio, near which earlier English immigrants had settled. Some were Hereford breeders, and what was more natural than that this young man from Herefordshire should cast his lot in the same direction? He acquired a few Whitefaces, became a successful exhibitor at local fairs and soon became recognized as a perceptive judge of beef cattle.

Morgan's reputation reached far afield with the result that in the middle 1870's he was employed as herdsman by T. L. Miller, Beecher, Illinois, probably the leading breeder and foremost Hereford advocate of his time. Under Morgan's handling, Miller's show herd scored remarkable successes and, at the same time, his talent as a top-rank judge became ever more widely recognized. His advice was sought by prospective buyers of Herefords, especially those who were relatively new to the business.

Among the latter was Charles M. Culbertson, a successful Chicago businessman who owned a fine farm at Newman, Illinois, and who had been recruited to the Hereford field by Miller. So when Culbertson in 1879 asked Miller to allow Morgan to go to Herefordshire to select a herd bull and some heifers for the new herd, Miller gladly extended permission. And, just as gladly, Morgan accepted the invitation to travel to his homeland.

He visited several herds in the rolling hills of Herefordshire, but nowhere did he see a herd-bull prospect that filled his eye so completely as did a two-year-old which he found at the Stocktonbury farm of T. J. Carwardine. This bull was Anxiety, first-prize winner as a yearling at the English Royal of 1877. Morgan's persistence led the owner to price the bull at the English equivalent of $1,000. Thus, Anxiety came to America to stamp his name upon the consciousness of American Hereford breeders for all time.

As a representative of Culbertson's show herd, Anxiety cut a wide swath at

*In recognition in the breed's homeland of Morgan's efforts in America, this loving cup was presented to him during an assembly at the Green Dragon Hotel in the city of Hereford, England. The inscription reads as follows: "Presented to George F. Morgan, Esq., by a few breeders in England, in recognition of his labors to establish the Herefords in America, July 25th, 1883." The trophy is now in the hands of Clyde C. Faler, Jr., a great-great grandson of Morgan.—Sotham photo.*

*This was the bull calf, Anxiety, as sketched by an artist some months before George F. Morgan "discovered" him in the Stocktonbury herd of T. J. Carwardine in Herefordshire, and purchased him for the new herd of C. M. Culbertson of Illinois. Anxiety was the founder, primarily through his son, Anxiety 4th, of the Hereford family which bears his name.*

virtually all of the leading fairs in the central states in 1879 and 1880. So deeply had he impressed all who saw him that his sudden and unexpected death following the 1880 show season precipitated a rush to buy his sons. Luckily Carwardine had used him lightly before allowing him to leave Stocktonbury, as a result of which a few Anxiety sons were available there. Among the first to see them were one of the Gudgell brothers and Thomas A. Simpson of the firm of Gudgell & Simpson, then of Pleasant Hill, Missouri.

They had been admonished before leaving home by the other Gudgell brother to seek out, and buy if possible, a "bull with an end on him," in which particular this youngster filled the bill, having inherited a characteristic which was predominant in the conformation of Anxiety himself. In fact, it was the impression Anxiety had left upon the consciousness of the Missourians which directed them to Stocktonbury when the course of events had made it apparent that only a few calves by Anxiety were going to be available in the neighboring state of Illinois. Three reasons accounted for this: 1, Culbertson's breeding herd was still small; 2, Anxiety had been so busily engaged on the show circuit that little time was left for herd duties, and, 3, the bull's premature death when only four years old.

Thus, George Morgan's contribution to American Hereford history becomes apparent.

But there is more. He was a member of the organizational committee which, in 1881, brought about the formation of the American Hereford Cattle Breeders' Association, now the American Hereford Association. Further, he was elected as a member of the first board of directors of the Association. His talents were recognized on both sides of the Atlantic, as was indicated when a group of leading English Hereford breeders at a meeting in the city of Hereford in 1883 presented him with an attractive silver cup as a testimonial to his extraordinary contributions to the breed's development and progress.

By the early 1880's, Morgan pulled up stakes in Illinois and headed westward to join the ambitious Alexander H. Swan in forming perhaps the greatest Hereford enterprise ever conceived—at least in the 19th century. With a vast acreage in southeastern Wyoming under control of the Swan Land & Cattle Company, Alex Swan was determined to be breeding 20,000 cows to purebred Hereford bulls by the end of his first five-year program, with numbers to go on up beyond that point.

Not only as a source of supply for the Swan outfit's own needs but also because he foresaw a great future for Herefords throughout the West, Alex Swan decided to establish a large registered Hereford herd, which he did in 1882 under the name of Wyoming Hereford Cattle and Land Association—later to become the Hereford Corporation of Wyoming, and still later the Wyoming Hereford Ranch. Headquarters were set up on Crow Creek, east of Cheyenne, and George Morgan was appointed general manager, considered at the time one of the most prestigious positions in the entire western cattle industry.

Morgan shortly was sent to England to buy foundation stock for the new herd, the result of which was an importation in 1883 of 146 head from eight of the leading English herds of the times. The pride of this shipment was Rudolph, acquired from the famous herd of Philip Turner of The Leen at a price of $3,500, an amount that topped all former record prices for British bulls. Hailed as "the mighty" Rudolph, in consequence of an impressive string of showyard triumphs, he died in 1885 before his worth as a sire could be well demonstrated. But he did draw wide and favorable attention to the breed.

Another large importation crossed the

*Morgan's influence in western cattle circles is attested by the fact that the advertisements of Wyoming Hereford Association at Cheyenne, to which he went from Illinois, bore his name as general manager and salesman. This one appeared in* The Breeders' Journal *in the middle 1880's, and emphasized the scale of this vast operation.*

Atlantic under Morgan's direction in 1884 to increase the Wyoming herd to 500 head of brood matrons. Plans for continued expansion were halted, however, first by drouth and a resultant shortage of range feed, and then by the savage winter of 1886-87, which decimated virtually every herd in the region and sent ranchers by the score to the wall. One of them was Alex Swan, who went bankrupt at the end of 1887.

This ended Morgan's connection with the operation but it is of interest to note that the herd which he guided in its formative years continued without interruption, but under successive ownerships, until the Wyoming Hereford Ranch dispersal in 1976. By that date it had been in continuous existence longer than any other substantial herd in American Hereford history. That final auction was held at virtually the same spot where the original English foundation animals first set foot nearly a century earlier.

Morgan then engaged in other ventures for a few years, some in the ranching field and some in mining activity, but eventually he returned to Herefords in 1898 when he assumed the management of the prominent herd of George H. Adams which, by then, had been moved to Linwood, Kansas, from its original home at Crestone, Colorado. When the owner's ill health brought about the dispersion of the Adams herd at record prices in 1903, Morgan immediately was named manager of the famous Weavergrace Hereford herd of T. F. B. Sotham at Chillicothe, Missouri. Ill when he arrived, he died some two weeks later, but his accomplishments during the preceding quarter-century make secure his place in Hereford history.

Perhaps as an epilogue, it is relevant to quote this comment of Alvin H. Sanders, early-day breed historian, who knew Morgan personally, and said of him in *The Story of the Herefords*:

"The name of George Morgan will ever stand conspicuous among those playing large parts in the introduction of Herefords in the western states. He was generally regarded as a keen judge of a good animal, and personally selected in Herefordshire some of the greatest cattle transferred to American soil during the period of extensive importations. In the course of his long career in the business he naturally acquired a great store of information concerning the breed on both sides of the water. His facility of expression, his aggressive personality and his keen sense of humor made him the life of almost any company of congenial spirits in which he might be found."

CHARLES GUDGELL

THOMAS A. SIMPSON

# Charles Gudgell
INDEPENDENCE, MISSOURI
1847-1916

# Thomas A. Simpson
INDEPENDENCE, MISSOURI
1822-1904

No Hereford herd in the history of the breed in America has come close to matching in sustained influence that which resulted from the constructive breeding activities of the firm of Gudgell & Simpson, first of Pleasant Hill, and later of Independence, Missouri. What this partnership, consisting of Charles Gudgell and his younger brother, James R. Gudgell, and their older partner, Thomas Alexander Simpson, accomplished during the herd's 40 years of existence, comprises a saga without parallel in American Hereford history, and perhaps in the history of any other beef breed in this country.

All three of these men were natives of Bath County, Kentucky. The father of the Gudgells, Joseph Gudgell, was one of that area's good livestock farmers, while Simpson, a distant cousin, had been a partner from time to time in some of the senior Gudgell's farming operations. Simpson was 25 years older than Charles Gudgell, who was two years older than his brother James. Simpson had gone in 1855 to Pleasant Hill, where he had operated a livery stable and traded in horses and mules until the border troubles led to his return to Kentucky a short time before the outbreak of the Civil War.

When illness overtook Joseph Gudgell in 1865, and the end appeared imminent, he asked Simpson to look after the sons, then 18 and 16 years of age. Perhaps as a mark of his willingness to assume responsibility for making decisions, Simpson years earlier had been nicknamed "governor", and this trait became clearly evident in his contribution to the partnership, especially during the firm's formative years.

Among Joseph Gudgell's assets at the time of his death was a tract of 480 acres of land located a few miles north of the village of Pleasant Hill, in west-central Missouri. The elder Gudgell had never seen this land, but it is interesting in speculating upon the course of Hereford history to wonder what this inheritance may have contributed to the decision of the sons to follow their mentor to Missouri. For it was not long after Simpson returned in 1870 to Pleasant Hill that his two young wards joined him in Missouri.

First they raised grade cattle, then briefly turned more or less in the direction of the then popular Shorthorn breed. In 1876 all three of the partners travelled to Philadelphia to attend the nation's Centennial Exposition, and there they saw the Hereford display which was forever afterward to affect the course of the breed's history. They studied especially the exhibits of T. L. Miller of Illinois, John Merryman of Maryland and H. C. Burleigh of Maine. They were particularly impressed with some of the Whitefaces that had originated in the herd of Frederick W. Stone of Guelph, Ontario, who had pioneered with Herefords in Canada in 1860. It was the Missourians' vision of what cattle like these might accomplish in the American West that led to a course of action.

The decision was made that Charles Gudgell should return home via Guelph, and, if he liked the cattle in the Stone herd as well as the descendants of Stone breeding seen at Philadelphia, should negotiate the purchase of some foundation stock. The result was the shipment of a bull and seven females to the Missouri farm. So favorable was the general impression created by these cattle that Charles Gudgell returned to the Stone farm the following year to buy another bull and several cows and heifers.

Also included in this 1877 shipment were a dozen young bulls which were grown out in Missouri and featured in an auction held in Kansas City in May, 1879, which Charles Gudgell later stated was the first public auction sale of purebred Herefords ever staged west of the state of Ohio.

These bulls brought an average price of $256 per head, mostly from men who had never seen Herefords before, and the sellers were so enthused over the reception accorded the bulls that they soon expanded their herd through

*Anxiety 4th, bred in Herefordshire by T. J. Carwardine of Stocktonbury farm, and imported by Gudgell & Simpson in 1881. In their herd he proceeded to gain lasting fame as "the father of American Herefords."–N. A. Throop sketch.*

female purchases made in Ohio from John Humphries and W. W. Aldrich.

These bulls and others of their own raising which they had sent west proved popular with ranchers but frequently were criticized on one significant point: their rear-quarters were too light. "Send us bulls that are not cat-hammed," urged the ranchmen. It was this complaint which led to Gudgell & Simpson's decision to go straight to the breed's fountainhead in the English county of Hereford to seek the means to improve the rear-quarters of their herd's output. They were aware of the strength of quarters of the bull Anxiety, which C. M. Culbertson had imported two years earlier, and which had been a first-prize winner at the leading Midwestern fairs in 1880. His impressiveness undoubtedly led James Gudgell and "Governor" Simpson to go directly to the Stocktonbury farm of T. J. Carwardine, Anxiety's breeder, near the village of Leominster, not far from the city of Hereford. There Simpson's eye soon fell upon the year-old bull named Anxiety 4th, and without really consulting James Gudgell, who never was an influential participant in the firm's purebred operations, he decided that this youngster was the bull upon which he would stake his recognized reputation for selecting "breeding" stock. James Gudgell's input in the firm was more toward representing the partners in their one-third ownership of a Colorado commercial cattle operation.

Thus the search for "a bull with an end" culminated in the importation of

1

2

3

**ON THE FACING PAGE:**

*In competition with the best Herefords of their time, the entries of Gudgell & Simpson acquitted themselves with great credit. All of these groups won first in their respective classes at the 1900 National Hereford Show in Kansas City, predecessor of the American Royal. Because of the possible historic interest in some of the animals pictured, they will be identified, left to right:* **Top**—*first-prize calf herd, consisting of Kandahar, Donna Ada, Bangle 5th, Bright Duchess 32d, and Bright Lass 16th.* **Center**—*first-prize young herd, consisting of Donald Dhu, Modesty, Mischief Maker, Honora 2d and Miss Caprice.* **Bottom**—*first-prize graded herd, consisting of Dandy Rex, Mischievous, Blanche 13th, Mischief Maker and Honora 2d.*

*It is worth noting that in this era cows that were shown also could be noteworthy producers of show cattle. Mischievous, in the bottom lineup, for example, was the dam of Mischief Maker and Miss Caprice, pictured here in their respective strings. It is of further interest that she later produced seven more calves, one of which was Mousel Bros.' famed herd bull, Beau Mischief.—All* Breeder's Gazette *photos.*

Anxiety 4th in a shipment of around 100 head to the Gudgell & Simpson farm in 1881. That same importation included Belle which, to the service of Anxiety 4th, produced Don Carlos, through which the Anxiety 4th influence largely descended, mainly through Don Carlos' sons, Beau Brummel and Lamplighter. A year earlier, Gudgell & Simpson had imported 61 head, and in 1882 another group of approximately 100 head from Herefordshire was added to their herd. These three shipments concluded Gudgell & Simpson's activity as importers.

As every student of Hereford history knows, Anxiety 4th's get not only met the requirements of the western rangemen but his descendants through successive generations revolutionized the Hereford breed in America. And the processes by which Gudgell & Simpson intensified his influence provided a whole new chapter on the technique of breed and bovine improvement.

In this endeavor, Charles Gudgell, the financial head of the firm and the final arbiter in all high-level decisions, collaborated with Simpson, a man recognized as possessing a real genius for forecasting the prospective usefulness of breeding animals. It was this perspicacity which enabled Simpson to get Anxiety 4th at a price of only $400. In fact, some of the English breeders urged him to reject this bull, which they thought might prove to be a disappointment in America, and thus reflect unfavorably upon English Herefords. But Simpson was firm in his conviction, and continued to play a significant role in the herd's operations almost until his death which occurred in 1904. James Gudgell had died in 1897, thus leaving Charles Gudgell to carry on, with the assistance of two sons, Charles and Frank, until the herd's final dispersal at auction in 1916. Gudgell himself died later in that same year at his retirement home in Pasadena, California, shortly after moving there.

Evidence is abundant that the improvement wrought in American Herefords by Gudgell & Simpson was the result of carefully thought-out plans persistently carried into execution. Charles Gudgell had studied the history of the origins of various breeds of domestic livestock and had been impressed with the fact that all of the great improvers of these breeds had practiced closebreeding. He concluded that it was by this process that those breeders had been able to fix in the breeds the desirable characteristics they wished to preserve and emphasize, and to eliminate the undesirable characteristics they wished to discard. Gudgell correctly reasoned that if this process had been effective a century or two before his time, it would continue to be equally applicable in this instance.

Anxiety 4th himself was the product of closebreeding, although it so happened that it was his individuality–his remarkably heavy hindquarters–that was the overriding factor in his purchase. But it obviously was not lost upon the Missourians that the sire and dam of Anxiety 4th were better than three-quarters brother and sister. Both were sired by a bull named Longhorns and the dam of each was a daughter of DeCote, both prominent names in English Hereford history. And there was yet another and more distant line of common ancestry.

As Gudgell & Simpson's breeding program proceeded, most of their fellow Hereford producers freely prophesied that the Missourians would ruin their herd by such intensive closebreeding. Their critics called it inbreeding, according to John M. Hazelton, former editor of *The American Hereford Journal* and author of the definitive history of the Anxiety 4th family. These observers, according to Hazelton, "pointed out the dire results that inevitably must follow a continuation of the practice." In deference to these opinions, the Missourians eventually did seek a suitable outcross . . . "but failing to find outside the herd cattle as good as they bred themselves, the practice of line-

*Portico of the Moreton Lodge farm home of Frederick W. Stone of Guelph, Ontario, the pioneer Canadian Hereford breeder. It was through these portals that Charles Gudgell passed to make the first registered Hereford purchases of Gudgell & Simpson. The old Stone home is gone, but the portico still stands on the campus of the University of Guelph, formerly the Ontario Agricultural College, which years ago engulfed the former Stone farm.–Photo by the author.*

*Most famous of the sons of Anxiety 4th was Don Carlos, the sire of both Beau Brummel and Lamplighter, progenitors of the lastingly notable Anxiety 4th lines of the 20th century. He also was a show bull of distinction, standing second, when almost seven years old, in the aged bull class of 10 head at the Chicago World's Fair in 1893 only to the imported Ancient Briton, then but 3½ years old. Beau Brummel ranked fifth in the two-year-old class in the same show, while Lamplighter was the first-prize yearling.—Lou Burke sketch.*

breeding was continued, with what beneficial results every breeder of Herefords today is familiar." Hazelton concluded: "The story of Gudgell & Simpson has no parallel in the history of American domestic livestock. To the student of that history, it is the most interesting and instructive chapter in the entire narrative."

It should be emphasized that Gudgell & Simpson combined with their closebreeding practices an almost unerring sense of selection. They culled ruthlessly from their breeding herd, sending to the range or slaughter all but those animals deemed worthy to carry on in their own herd or other select breeding herds. Charles Gudgell probably should be credited with recognizing clearly that the herd's program of blood concentration would intensify the inferior characteristics as well as the good ones, and hence sought consistently to discard those animals of their production which were regarded as potentially harmful. The result was a concentration in the Gudgell & Simpson herd of the superior characteristics of the Anxiety 4th family, and a diminution of the others, this being the key to the family's prepotency for so many generations as an improving force in the Hereford breed.

In a statement in a 1900 sale catalogue, Gudgell & Simpson made the only known commentary in defense of their breeding practices. Signed with the firm's name, it probably represented the thinking and convictions of Charles Gudgell. Excerpts follow:

"Perhaps some who have not given a degree of study to the well-established principles that govern breeding for physical improvement will at once say 'inbred' if a sire appears more than once in a pedigree, and from prejudice alone condemn the practice without any further consideration, although it should be patent to anyone that our operations in this direction have been conducted upon a very conservative scale. And yet inbreeding is the course that has been pursued by all the breeders of improved livestock who have effected any marked improvement in their specialty. . . .

"In our own breeding operations we have not deemed it necessary or wise to combine such close affinities as was the ruling practice with these early masters of the art [the patriarch Jacob and the great modern British improver, Robert Bakewell], contenting ourselves, as an extreme measure, with the mating of two animals by the same sire, but with dams of entirely different strains of blood. In this way we got our outcrosses through dams instead of through sires as is the practice of most breeders. . . .

"We realize that in our course of breeding we have to encounter the prejudice engendered by the fact that a certain once popular and valuable strain of cattle was ruined in constitution and otherwise by inbreeding. This sample of injudicious breeding that is so often cited as proof condemnatory of the practice of linebreeding is, on the contrary, a recommendation of linebreeding in itself. The breeders of this strain of cattle who bred them with an eye single to their ideal of pedigree perfection realized their object in the

**GUDGELL & SIMPSON,**

**INDEPENDENCE, MO.,**

**Breeders of Pure-Bred Hereford Cattle.**

**The largest and one of the oldest herds in the Central West. Particular attention invited to the uniformly high character of our cows and heifers—all of which are of our own breeding.**

Chief stock sires now in service: ***DON CARLOS 33734,*** second-prize aged bull at Columbian Exposition, and his sons ***LAMPLIGHTER 51834***—first-prize yearling at World's Fair—***BEAU BRUMMEL 51817, DRUID 46833*** and ***SPARTACUS 51842.***

Inspection invited. Independence is but 10 miles from Kansas City and hence easily accessible from all points.

*This was the Gudgell & Simpson advertisement which appeared weekly in* The Breeder's Gazette *at around the turn of the century. Names in the herd-bull battery deserve special note.*

same measure as did other linebreeders who bred only for physical perfections–the one secured a pedigree unsullied and the other a physique unsurpassed. . . .

"As regards our own breeding operations, we can only say that we have more than preserved the size and have fully maintained the constitutional vigor and feeding qualities of our original stock, and have, at the same time, evolved in our young things a very material modification and refinement of the head, horns and neck, as well as thickened and straightened up the hindquarters to a degree that has not been surpassed, if at all equalled, by any other modern breeder. As a result of having a well-defined object in view and the pursuit of that object in the lines we have followed for nearly two decades, we have developed a type of Herefords with characteristics that mark them from others. . . ."

The following condensation of the principles expressed above appeared in the firm's advertisements: "Uniform in type and representing a concentration of blood that insures transmission of their thoroughly-fixed character."

The solidity of the foundation upon which Gudgell & Simpson built is apparent in the fact that virtually all American Herefords 100 years after the birth of Anxiety 4th are descended from him–most of them tracing to him in many lines of their ancestry. Rare indeed is the Hereford of the final decades of the 20th century which does not have as an ancestor such descendants of Anxiety 4th as Don Carlos, Beau Brummel, Lamplighter, Prince Domino, Beau Mischief and scores of others of similar descent, right down to the Line 1's, RC Mischiefs and others which bask in the limelight of the 1980's.

In addition to his personal involvement in herd operations, Charles Gudgell was a prime mover in the formative years of the American Hereford Cattle Breeders' Association. He was a member of the first board of directors, named in June, 1881, served for several years on the executive committee of three, was elected secretary and treasurer in January, 1884, continuing in this capacity until 1887, and served the Association as its president in 1906.

Through his constructive activity as a breeder and his dedicated service to the Association, Charles Gudgell was a genuine benefactor of the Hereford breed and industry. Indeed, the breadth of the recognition accorded him as an authoritative spokesman for the breed is shown in the fact that he wrote the Hereford section of the definitive *Cyclopedia of American Agriculture*, published in 1908.

But as important as Charles Gudgell's role was, the significance of Simpson's contribution to the firm must not for a moment be minimized. After all, the decision to import Anxiety 4th was his, as were many of the matings which made the herd lastingly famous.

# Alexander H. Swan

CHEYENNE, WYOMING

1831-1905

The most spectacular operator in the Hereford field in the heyday of his career was Alexander Hamilton Swan. Born in Pennsylvania in 1831 and reared on a farm in that state, he went in 1855 to Ohio where he lived for seven years before moving on to Indianola, Iowa, shortly after the Civil War. It was there that his extraordinary flair for promotion first came widely into view.

Starting in business at Indianola as a grocer, he soon developed a genius for speculation. One of his first schemes was in connection with building a railroad to Des Moines, some 20 or 25 miles to the north. Then he began to speculate in Iowa farmland, acquiring many hundreds of acres, which led to extensive cattle-feeding activities. When a market for bulls for the western range began to materialize, some 2,500 acres of Swan's Iowa holdings were changed into a breeding establishment with a cow herd of some 600 head. Other enterprises in and around Indianola with which he was involved included a coal mine, brickyard, flour milling, a cannery and a bank or two.

But as extensive as these activities were—involving visits from English capitalists and other investors—Swan still yearned for bigger things and the "more perfect freedom" of the new country farther west. Having great ambitions to become even more heavily involved in the cattle business, he migrated to Wyoming in the early 1870's, when well past the age of 40.

Operating first as Swan Bros., at Cheyenne, Wyoming Territory, in conjunction with his brothers, Henry and Thomas Swan, the first volume of the American Hereford Record, issued in 1880, lists the Swans as the owners of 83 registered Herefords, including both bulls and females.

In the spring of 1878, Alex Swan paid T. L. Miller of Beecher, Illinois, $10,000 for 40 bulls, which is considered the first substantial introduction of Hereford blood to the Wyoming range. A second purchase of 50 head came shortly thereafter. These, of course, also went into range use.

Intent upon further expansion, Alex Swan, as the firm's dominant partner, plunged on. Having found Scottish venture capital, he in the spring of 1883 organized the Swan Land and Cattle Company in Edinburgh. He became general manager. Controlling the Two Bar, Double O, Horse Creek, Kingman and other properties, covering a range 50 by 150 miles in extent, the combined herds under the corporation's ownership totaled 120,000 head. As an industry leader, he became the second president of the Wyoming Stock Growers Association in 1876, serving in that capacity until 1881.

At the outset some 5,000 cows were being bred to purebred Hereford bulls, with an increase to 10,000 head a year scheduled to be reached within a year, and with the total to be expanded to 20,000 such matings within five years. Realizing that fenced pastures were necessary to the success of the herd-improvement program he envisioned, Swan became a pioneer in fencing on the northern plains. By the early to mid-1880's some 100,000 acres were under fence, and wire was being shipped in by the railroad carload.

*On the way in his search for ranching fame in Wyoming, Alexander H. Swan lived for a time at Indianola, Iowa, and proved to be an ambitious and innovative operator in several lines, including cattle production and feeding. This was the old Swan farm house at Indianola.*—Breeder's Gazette *photo.*

Aggressive promoter that he was, Alexander Swan sensed a great commercial opportunity in the growing demand on the range for Hereford bulls. His vision of popularizing this breed in the West led to his formation in 1883 of the Wyoming Hereford Cattle and Land Association (usually shortened to Wyoming Hereford Association), which in actuality was not an association but a separate company which had no connection with the Swan Land and Cattle Company except that some stockholders held stock in each. The ranch upon which the registered Hereford herd was to be carried comprised some 30,000 acres on Crow Creek, a few miles east of Cheyenne. Swan was president and George F. Morgan, a native of Herefordshire, who had played an important role in the pioneering work done earlier by T. L. Miller and C. M. Culbertson in Illinois, was engaged as general manager and salesman.

Without delay Morgan arranged a number of substantial importations, including the entire 200-head herd of J. H. Yeomans of Stretton Court farm, then among England's leaders. Another large importation was made in 1884. One of these early importations was headed by Rudolph, selected by Morgan from the famous herd of Philip Turner of The Leen at a then record price of $3,500. The 2,600-pound Rudolph—the "mighty" Rudolph as he was widely publicized—became a sensational winner at the leading fairs of 1883-84 and an offer of $5,000 was reportedly refused for him. Although he died prematurely in 1885 as a result of an injury, his sturdy conformation and impressive style were reputed to have won, among the thousands who viewed him, many converts to the Hereford cause. And his reputation was further enhanced when a steer sired by him, Rudolph Jr., was acclaimed sweepstakes winner at the American Fat Stock Show in Chicago and hailed in these words: "From the standpoint of the butcher, he was the best steer so far the winner of this premium."

By 1885 the Wyoming Hereford Association, under Swan's direction, was

*The Wyoming Hereford Association, headed by Alexander Swan, in 1882 paid $3,500 for the English-bred Rudolph. Selected by George F. Morgan, general manager of the Wyoming herd, he headed an importation of 146 head which arrived in America early in 1883.*—The Breeders' Journal *photo.*

*These cows and calves on the property of the Wyoming Hereford Association are samples of the product of the foundation laid by Alexander Swan when he embarked upon his ambitious program in the early 1880's to build the West's largest and best Hereford herd.*—Breeder's Gazette *photo.*

advertising in *The Breeders' Journal*, then the Hereford industry's principal public advocate, that it was the "largest herd in the world," then "numbering over 500 head." The advertisement continued: "At its head stands the first-premium bull, Rudolph, supported by the best Hereford bulls that could be selected in England." Continuing, the advertisement stated: "For range purposes our stock cannot be equalled. They are bred on the range, and do not need to pass through any acclimating process. We invite the attention, and shall be glad to show our stock. Prices reasonable."

Observers were predicting that the time soon would come when this concern would be selling 10,000 bulls a year.

Alas, that time never came. Triggered perhaps by the catastrophic winter storms of 1886-87, financial problems arose. In addition to enormous death losses in the Swan grade herd caused by a succession of blizzards of unprecedented intensity, an over-extension of credit, a sharply declining cattle market and a degree of miscalculation of liquid resources sent the Wyoming Hereford Association and the other Swan interests into receivership late in 1887. He played for big stakes and lost.

But during its period of activity, the Wyoming Hereford Association sold many, many hundreds of purebred Hereford bulls which spread the fame of the Whitefaces throughout the northern range country. When the home-raised supply was exhausted, bulls in large volume were brought in from the Central states. For example, a single shipment in the spring of 1886 included 17 carloads of grade Herefords. Most of the more enterprising ranchmen of Montana and Wyoming, along with some in adjacent regions, thus had recourse to a source of supply of bulls that placed within reach of the cattle industry of the northern plains blood that left its mark for many a year. The Hereford leadership thus established in that region continued on and on.

It was a fortuitous circumstance that the herd which exerted such an early and potent influence upon range production in the West was not immediately scattered to the four winds. It was continued, in fact, without interruption under successive ownerships, until a final dispersal in 1976. The property then was the Wyoming Hereford Ranch, as the entity became known with a change of ownership in 1921.

The ranch, however, was not long without a registered Hereford herd, although on a considerably reduced scale. Sloan and Anna Marie Hales of Cheyenne, shortly after the 1976 dispersal, acquired the headquarters unit which comprised a portion of the property where Alex Swan started his purebred operations, and as of 1981 Wyoming Hereford Ranch again was a going concern, carrying on the name and traditions of almost a century.

It may be of interest also to point out that Alexander Swan was one of the founders of the stockyards at Council Bluffs, Iowa, and later was largely responsible for the establishment of the Union Stock Yards in South Omaha, Nebraska. He also was president of the South Omaha Land Syndicate. The latter concern purchased a considerable area of land, subdivided it and started the town of South Omaha, which became a fast-growing municipality. He served in the state legislatures of both Iowa and Wyoming.

That Swan was one of the all-time great promoters of the cattle kingdom there can be no doubt. That the Hereford breed was a major beneficiary of his enterprise is fully evident. Some who were involved in trying to bring a sense of order to the financial chaos which he left behind when his financial "ship" sank called him a reckless plunger, but others sympathetically felt that his ambitions simply outran his judgment. One of these was an employee who worked in the Swan office for several years. Decades after the financial debacle, he remembered Alexander Swan "as looking like Uncle Sam—lank, gaunt, a long face, a beard, gray hair, knock-knees, and fine blue eyes; a personality whose failure was greatly regretted."

Despite his great personal misfortunes, the Hereford industry is indebted to him not only for his enterprise in scattering thousands of white-faced bulls across a vast expanse of range country, and thus bringing the breed to favorable attention in the West, but also for his leadership role in the breed association, which he served for one term as its fifth president.

# William S. Van Natta

FOWLER, INDIANA
1830-1911

Few breeders in the developing years of Herefords in the United States exerted a greater constructive influence than William S. Van Natta of Fowler, Indiana. A practical stockman who knew all there was to be known about the grazing, feeding and marketing of beef cattle, he was fortunate in forming a working partnership at the start of his undertaking with purebred Herefords with Moses Fowler, a banker who owned some 25,000 acres of virgin grassland in Indiana's Benton County. Together they set about the creation of a herd of Herefords that would rank with the breed's leaders, and together they succeeded in doing that very thing. Fowler, who once had been associated with C. M. Culbertson in the packing business, had the money, but it was Van Natta who had the practical working knowledge that made their Hereford enterprise a success.

Fowler and Van Natta formed their partnership in 1877, and during the following year the foundation for their herd was laid with the purchase of some 25 cows and heifers from the enterprising T. L. Miller of Beecher, Illinois. One of these later became the dam of the bull Fowler, one of the most noted of the early products of the herd. Fowler's sire was the T. J. Carwardine-bred Tregrehan, an English Royal show winner in his homeland and as a member of the Fowler & Van Natta show herd the "best bull of any age or breed" at some of the leading mid-western shows on this side of the Atlantic in the early 1880's.

Tregrehan, whose sire was a half-brother to Anxiety which came to America earlier, proved to be a great success as a breeding bull. Although said not to have been a big bull, his beefiness, according to available accounts, was remarkable. However, he had a deep dimple in his broad and straight top-line to which some breeders objected. Van Natta's rejoinder was: "Well, if he didn't carry a great deal of meat on his back, he would not have a dimple."

By the early 1880's Fowler and Van Natta were carrying up to 2,500 head of steers on their pastures, in addition to the purebred breeding herd, and were well pleased to find that the grade Hereford steers not only met their expectations as grazers, but did even better than anticipated. Furthermore, they learned very shortly that the Hereford beeves responded exceedingly well to the feeds afforded them in the finishing ration, and that they met favor with the packers when they went to market.

With this reassurance of Hereford beef-making performance, Fowler & Van Natta went steadily forward in expanding and improving the herd of purebreds on what became widely known as Hickory Grove Farm. Winning steers were exhibited as early as 1882 at the Chicago Fat Stock Show, and the "grand sweepstakes for best two-year-old steer, all breeds competing," at the 1883 Kansas City Fat Stock Show was a Hickory Grove representative.

That same year, 1882, breeding cattle entries represented the herd at five Indiana, Illinois and Missouri fairs, with Tregrehan winning the sweepstakes award for the best bull of any age or breed at two of them, and ranking second in his class at the other three, including the Great St. Louis Fair.

Tregrehan's son, Fowler, by the middle 1880's had gained general acknowledgement as one of the best bulls of any beef breed in America at that date. As a three-year-old in

*William S. Van Natta stood here in front of his home at Fowler, Indiana, not far from where he was born in 1830.*—Breeder's Gazette *photo.*

*In an era when mature individuals were among the features of the leading show herds, this was Van Natta's undefeated aged herd of the 1908 season, all products of his Prime Lad family of Herefords. At the head of the line was Prime Lad 9th, purple-ribbon winner at the American Royal and International, and one of the most noted winners of his time.*

1886 he won repeatedly, and headed the Hickory Grove entry which won the sweepstakes award for bull and three of his get, all breeds competing, at Kentucky's heralded Shelbyville fair, a high point during the Hereford breed's celebrated Hereford "invasion" of the Kentucky stronghold of the Shorthorns. Of still greater importance to his owners, he proved to be a wonderful sire and contributed in full measure to winning wide acceptance for Herefords among American stockmen in the final years of the 1800's.

Van Natta was at the helm of the firm's merchandising endeavor which initiated a series of annual public auctions as early as the middle 1880's. The first annual catalogue, listing with their pedigrees all of the animals in the herd, including stock bulls, breeding matrons and all of the younger animals–bulls and heifers–was published and widely distributed.

The firm's emphasis upon the practical end of the business was stressed in the introduction included in the catalogue of an auction staged at Dexter Park, in Chicago, on November 13, 1885. It read: "It is sufficient to say that the cattle offered in this sale are straight in every particular. There are no barren animals or doubtful breeders. All those that have not young calves at foot (except a few that are too young) are safe in calf to first-class bulls, and in good breeding condition, having been kept on grass and hay alone. This will be a rare opportunity for parties who wish to make money."

The following spring an auction was held in Kansas City, and a year later, on May 10, 1887, the Hickory Grove Farm was the site of an auction in which, again, the emphasis in the announcement was upon elements of likely interest and concern to prospective purchasers, in such words as these:

"We aim to treat all our cattle so they will give satisfaction to their owners, and do not believe in over-feeding and pampering the animals. The cattle are perfectly healthy, and can be shipped anywhere in the United States. . . . We want it distinctly understood that we will have no sham bidders, and if the people want the cattle it is only necessary to bid to get them." Another point, as emphasized in the herd's advertisements, was this: "Custom for years to select and sell to butchers every fall all plain cattle in the herd."

Thus was confidence in the herd, and in the breed, built.

On the death of Moses Fowler, the herd was continued by Van Natta and his son, Frank, under the firm name of W. S. Van Natta & Son. Its success continued unabated. Like most of his contemporaries except Charles Gudgell and T. A. Simpson of Missouri, Van Natta frowned on heavy and continuous concentration of the same bloodlines. In his search for a new outcross he attended a sale of richly bred imported bulls staged by C. S. Cross at Sunny Slope Farm, Emporia, Kansas, in 1898 and bought there at a price of $1,000 the yearling March On, a product of some of England's best bloodlines. March On sired for the Van Nattas a procession of winners and the sires and dams of other winners and top breeding cattle for other breeders.

Then for use in a cow herd representing this admixture of bloodlines came Prime Lad, and if earlier bulls had lent substance to the stature of Fowler & Van Natta as breed pioneers, the reputation of W. S. Van Natta & Son as breed builders began with Prime Lad. For them this descendent of Anxiety on his sire's side and out of an imported cow of top English bloodlines became one of the most successful show bulls of his time, but of greater importance he founded a significant family of Prime Lads.

How the herd ranked among its contemporaries in the showring is evident in the fact that not only did Prime Lad cap an illustrious career as a show bull by winning the senior and grand championship at the St. Louis World's Fair

*In one of the most prestigious Hereford shows of its time, Prime Lad was senior and grand champion bull for W. S. Van Natta & Son at the St. Louis World's Fair of 1904, more formally known as the Louisiana Purchase Exposition. Further, the Van Nattas also showed the senior and grand champion female, Lorna Doone. How well these two bred on was demonstrated by the fact that Prime Lad sired the noted winner Prime Lad 9th, while Lorna Doone was his maternal granddam.*

in 1904 but another Van Natta entry, Lorna Doone, sired by the Van Natta-bred Christopher, also of Anxiety descent, became the senior and grand champion female there. This performance lent added prestige to the herd in an event in which representatives of most of the period's leading herds competed, but the show itself also gave great impetus to the Hereford advance into the 20th century. So the Van Natta contribution is evident.

Prime Lad sons gained wide prominence through their success as sires, not only in the home herd but also many others. One of these, Prime Lad 16th, whose dam was the champion Lorna Doone, sired Gay Lad 6th and Gay Lad 16th, which became the progenitors of two widely recognized families of Gay Lads–one in Missouri for Overton Harris and the other in Texas for C. M. Largent.

William S. Van Natta, a man who liked people and was recognized for a fine sense of humor, also was in the vanguard of those pioneers who had the foresight to organize a national breed association. In fact, he was the first chairman of the Herd Book Committee of the American Hereford Cattle Breeders' Association, as the national breed organization was known at its inception. Then he became the association's second president.

When W. S. Van Natta's life ended in 1911, he had come far from his Indiana log-cabin birthplace of some 81 years earlier. And his accomplishments as a Hereford breeder of unswerving dedication and tenacity were widely recognized. Of him, *The American Hereford Journal* said: "As a Hereford breeder he had few equals. His herd was one of the oldest in the country, and was claimed by many good judges to be the best they ever saw." Alvin H. Sanders, livestock historian, editor of *The Breeder's Gazette*, and a personal friend, wrote: "His career as one of the greatest cattle breeders this country has yet produced abounds in inspiration for those who would follow in his footsteps."

Of incidental interest to latter-day Hereford breeders is the fact that a nephew, J. W. (Jack) Van Natta of Battleground, Indiana, kept alive the family name in Hereford circles through almost all of the subsequent half-century, and followed in the footsteps of his uncle when he was elected president of the American Hereford Association in 1959.

ADAMS EARL

CHARLES B. STUART

# ADAMS EARL
LAFAYETTE, INDIANA
1819-1898

# CHARLES B. STUART
LAFAYETTE, INDIANA
1851-1899

No early-day Hereford-breeding establishment had a more auspicious beginning than did Shadeland Farm, located a few miles southwest of Lafayette, Indiana. Owned by Adams Earl, a prosperous merchant and banker in that city, with interests also in railroad-building and meat-packing, its 1,400 acres of fertile farmland were devoted to agriculture and cattle feeding, and it was the latter activity which persuaded him to cast his lot firmly with the Whitefaces. He noted first that some neighbors were topping the Chicago market frequently with white-faced cattle and when he experimented with a few on his own place, where they mingled with steers of variegated colors, he very shortly satisfied himself that the Hereford breed was destined to succeed.

Finding few animals of the caliber he desired available in American herds, he went to England in the summer of 1880 seeking herd-foundation stock of the top rank. In this endeavor he was assisted by his son-in-law, Charles B.

Stuart, a capable young attorney, who was to become his partner at Shadeland which was to gain fame as the great American Hereford showplace of its time. The few modest purchases made on this 1880 trip constituted the foundation of the Shadeland herd. This acquisition merely whetted the Indianans' appetite for more and even better Herefords.

Not unmindful of the importance of establishing the standing of the Herefords in the commercial beef-making industry, and well aware that it was steer performance which had turned Shadeland to Herefords, the Indianans gave much attention to the development of show steers. The steer named Wabash, said to have been the only imported steer ever to win at the then great Chicago Fat Stock Show, predecessor of the International, was exhibited two years in succession. He was champion two-year-old in 1882 and reserve champion in 1883. It is recorded that as a three-year-old "he weighed 2,350 pounds and when slaughtered, netted 70.35 per cent" up to that time a record dressing figure.

One of the outstanding exhibits of 1888–after Shadeland was well established as a Hereford mecca–was a group of steers by a Shadeland herd bull and out of grade cows which, after winning first, second, third and fourth prizes as individuals, won the Marshall Field trophy for the best exhibit of fat cattle. Such successes, they were confident, helped greatly in "selling" Herefords to commercial stockmen. Furthermore, during all of these years, Adams Earl was running a large herd of grade cattle, further effectively demonstrating the utility and superiority of purebred Hereford sires in practical beef-making operations.

Almost simultaneously with their beginning years in exhibiting prime Hereford steers, Adams Earl and Charles Stuart became so greatly impressed by the showring successes in Hereford breeding-cattle competition of the entries of Tom Clark that they commissioned him to return to his native Herefordshire to buy for them a large selection of the very best that English breeders could be persuaded to sell. At that time, Clark was located at Elyria,

*A product of the noted Herefordshire herd of John Price, Garfield headed a sensational importation of Earl & Stuart made during the summer of 1882. The Perfection, Woodford, Fairfax and other once-prominent lines were descended straight from him.*

Ohio, and during subsequent years added to his reputation by successful Hereford operations at Beecher, Illinois.

Clark took plenty of time in inspecting most of the then-prominent herds in the breed's homeland and when, in the spring of 1882, he assembled his selections for export to their new home in Indiana, it was a superb lot of the breed's best. In fact, it was generally agreed by both press and public that no such collection of cattle had ever before left Herefordshire.

It included a long line of prize-winners at the English Royal as well as a remarkable collection of top breeding-herd prospects, and it was reported that no fewer than 50 leading English Hereford breeders gathered to witness the departure of the shipment. Matching the enthusiasm of the English sellers was the air of high expectancy in America as breeders on this side of the Atlantic awaited the arrival of the 125 head shipment which was expected to give unparalleled impetus to breed progress in the New World.

Heading the importation was the flashy yearling bull, Sir Bartle Frere, first-prize winner at the Royal show of 1881, for which $3,000 was paid, a record price for a Hereford bull in England up to that date. Interestingly enough, he was bred and sold by the same T. J. Carwardine of Stocktonbury fame who had bred and sold Anxiety to C. M. Culbertson and Anxiety 4th to Gudgell & Simpson, but Sir Bartle Frere was not closely related to the Anxiety line. His sire was Lord Wilton, an unrelated bull much more highly regarded by Carwardine than the Anxietys, while his dam was a daughter of Longhorns, which sired both the sire and dam of Anxiety 4th. Sir Bartle Frere matured into an impressive individual, drawing much attention to the fledgling breed as he won numerous first-prize honors for Earl & Stuart's show string, and then as a sire in the breeding herd, especially of herd-building heifers.

But it was another bull, Garfield, bred in the herd of John Price of Court House farm and descended from such breed-foundation sires as Sir Thomas 20 and Sir Benjamin 36, that proved to be the key Earl & Stuart acquisition. Almost a year younger than Sir Bartle Frere, he arrived in America a few months later after having made the rounds of the 1882 English shows with remarkable success. In a class of crack yearlings, and before six different sets of judges, he was awarded the first prize without exception in that string of English shows, a career without parallel up to that time in the history of the breed in England. These successes included the English Royal, where he was referred to as "the phenomenal yearling," and was pronounced the "best specimen of early maturity of any age or breed."

Not as stylish a bull as Sir Bartle Frere, but possessing greater substance, his breeding performance met his owners' every expectation. He fully deserved the fame which came to him as a sire, the results of his matings with daughters of Sir Bartle Frere being especially outstanding.

After Garfield's arrival at Shadeland no show string was taken out to the fairs until 1886. This was the year of the celebrated Hereford "invasion" of Kentucky, which previously had been the stronghold of the Shorthorns. At Shelbyville, a son, Earl of Shadeland 9th, won the sweepstakes award as the best Hereford bull of any age, and at Lexington he was awarded second place in a ring for bulls of any breed under two years old—this in spite of the threat by one of the managers of the fair that "the Whitefaces would be sent back across the river [the Ohio] without a ribbon."

This winner was but one of the famous line of Earls of Shadeland, all sired by Garfield, which contributed immeasurably in popularizing the Hereford breed to cattle growers and feeders during the last decade or two of the 19th century. This was an important formative period for the Whitefaces, and Shadeland Farm, under the management of John Lewis, who came from Herefordshire to assume the post, became the showplace of the breed in the United States, and as such contributed immensely to the advancement of Hereford interests generally.

Nowhere else were such pains taken to display the cattle to visitors and prospective buyers in the most attractive manner possible. Carefully sorted and grouped by ages: the calves, the yearling bulls, the yearling and two-year-old heifers, the brood matrons, the breeding bulls in their paddocks, the show herd in preparation—all presented a picture that never failed to make a deep impression on visitors.

It was a few years after the coming of John Lewis that another famous bull, The Grove 3rd, was acquired from C. M. Culbertson, who had imported him

*One of the most famous of Hereford establishments was the Shadeland Farm where a noted herd was operated by Adams Earl and Charles B. Stuart. Alvin H. Sanders of* The Breeder's Gazette *once recalled the "bull shows" staged in the barn lots by Stuart and John Lewis, the Shadeland manager.*

from England. The caliber of this sire was indicated by the fact that $7,000 was paid for him when he was 11 years old.

But it was the Earls of Shadeland, primarily, which left the Earl & Stuart imprint on the Herefords of the latter 19th century and the early years of the 20th. In addition to "the 9th," the leaders among them included the Earls of Shadeland 12th, 22d, 30th, 41st and 47th. As a show bull, Earl of Shadeland 22d was a sensational winner in the latter 1880's, historians relating that he "won every prize for which he showed." Almost equally remarkable for his showring exploits was Earl of Shadeland 30th. But it was Earl of Shadeland 41st which left the most lasting imprint on the breed as a sire. This was mainly by virtue of the fact that he sired Columbus, the sire of Dale, which, in turn, sired Perfection. The latter bull's sons included two whose names were famous in Hereford circles well into the second quarter of the 20th century. One of these was Perfection Fairfax, noted sire of winners for W. T. McCray of Indiana, and the other was Woodford, which brought fame as a Hereford breeder to the name of Kentuckian E. H. Taylor, Jr.

Despite the immense contribution made by Earl & Stuart in establishing the superiority of the Hereford breed in America, this may indicate the possibility that the product of their effort might have had greater long-range influence upon the breed had the owners resorted to a program of closebreeding to Garfield rather than diffusing his genetic potential through the outcrossing program to which all of the leading breeders of the times, other than Gudgell & Simpson, were wedded. But in the years until 1900 when the herd was sold, following the death of Adams Earl in 1898 and that of Charles Stuart in 1899, no herd wielded greater immediate influence upon the breed and its destiny than did this one.

Adams Earl was active in breed councils throughout the years of his connection with the Hereford industry, was one of the organizers of the breed association and was its president for the 1890 term. Years earlier he had served as the association's first treasurer. And Charles B. Stuart is credited by early breed historians with having been the framer of the association's rules and by-laws. Further, he was a leading member of its executive committee probably for a longer period than any other individual.

# Charles Goodnight

PALODURO and GOODNIGHT, TEXAS
1836-1929

Great impetus was given to Hereford expansion into the Southwestern range country by the pioneer ranchmen, the acknowledged leader of whom was Charles Goodnight. A Macoupin County, Illinois, boy who became an adopted Texan at the age of nine, he remembered years later having seen buffalo ranging along the Trinity River within two miles of the site of present downtown Dallas.

A restless lad who did not take to the farming activities in which his stepfather was engaged in the Lone Star State's Milam County, he shortly struck out on his own. According to his biographer, J. Evetts Haley, "he was hunting with the Caddo Indians beyond the frontier at thirteen, launching into the cattle business at twenty, guiding Texas Rangers at twenty-four, blazing cattle trails two thousand miles in length at thirty, establishing a ranch three hundred miles beyond the frontier at forty, and at forty-five dominating nearly twenty million acres of range country in the interests of order. At sixty he was recognized as possibly the greatest scientific breeder of range cattle in the west, and at ninety he was an active international authority on the economics of the range industry."

Alvin H. Sanders, editor of the *The Breeder's Gazette* and leading livestock historian of the early 20th century, wrote in 1914 that "it seems to be generally allowed that the credit for the revolutionizing of the blood of the Texas Panhandle cattle along Hereford lines is largely due to Charles Goodnight." He first entered into cattle-breeding in 1856 in Palo Pinto County, Texas, and early set about to improve the herd which he was handling on shares. The only way open to him was through selection, that is, picking the best raised in the herd for future breeding purposes, but by this primitive means he not only markedly improved the standard of the herd but more importantly learned well an important lesson in how herd improvement is wrought which was to be the hallmark of his activities through later years.

In 1866 he laid out what became known as the Goodnight Trail to Fort Sumner, New Mexico, and later blazed an extension, known as the Goodnight-Loving Trail, all the way to Cheyenne, Wyoming. He prospered to the extent that he and his wife, to whom he was wed in 1871, decided to buy two large tracts of Colorado land, one in the vicinity of Pueblo and the other 40 miles from Trinidad. He engaged not only in cattle raising there but also undertook some rather extensive farming operations. He also branched out in the banking business, just in time for the nation's financial panic of the early 1870's to wipe him out with the exception of about 2,000 cattle.

His thoughts then turned back to Texas, and especially to a specific portion of the Panhandle which he had seen while scouting for the Texas Rangers. This was a vast canyon area which broke the plains a dozen miles or so southeast of the present city of Amarillo. Watered by the Prairie Dog Town Fork of the Red River, this normally small stream had cut a system of canyons nearly 60 miles long. The canyon walls ranged up to almost 1,000 feet in depth, while the canyon itself varied in width from a few hundred yards to extensive badlands 15 miles across. The high bluffs were more effective than posts and wire in fencing the animals, and the water and grass were sufficient for many thousands of cattle. Having made his decision to make a new start in Texas, Goodnight in 1876 loaded the family possessions and assembled his cattle for the journey to what became widely known as Palo Duro Canyon.

Upon arrival it proved to be impossible

*Col. Charles Goodnight with Mrs. John Adair of the historic JA Ranch in 1917, in a picture taken long after the pioneering times in the Palo Duro Canyon.—Photo courtesy of White Deer Land Museum, Pampa, Texas.*

*The commodious residence which Colonel Goodnight built at Goodnight, Texas, looked in 1976 very much as it did in the beginning except that the small trees visible here had grown large and tall.*

for the wagons to follow the cattle into the canyon. The terrain was too rough. Goodnight therefore took the wagons to the rim, unloaded them and took them apart, and with ropes lowered the wagon parts and the supplies down into the great cut. There, they were reassembled and the drive to the projected headquarters site continued. Here, as was pointed out in the C. L. Douglas book, *Cattle Kings of Texas*, at a location some 200 miles from the nearest outpost, and with his herd of 2,000 head salvaged from the panic, was the birthplace of one of the great ranches in the history of the Southwest.

After getting what came to be known as the Home Ranch established, with its house and corrals, Goodnight in 1877 made what proved to a propitious return to Pueblo to conclude some unfinished business. For while there he formed a partnership with John G. Adair, a wealthy investor from Ireland who had succumbed to the lure of the West and the promise of good returns from cattle ranching. Adair put up nearly a half-million dollars, an immense sum for those times, and the ranch took its name from the initials of his first and last names–JA. Goodnight bought the Palo Duro Canyon area and surrounding territory until the JA became one of the most famous of the cattle kingdoms–more than 1,335,000 acres with over 100,000 cattle at its zenith. The count was 63,000 head when the partnership was dissolved in 1888.

It was inevitable that he should become an industry leader, which he did by founding the Panhandle Livestock Association. He also sought to prevent drovers from crossing members' ranges, which was not completely successful. It is also said that Goodnight devised the first chuckwagon.

The Adair partnership enabled Goodnight to pursue the means for improving the JA herd as he had envisioned years earlier when he was running his first small operation in Palo Pinto County. He did not like the Herefords he had first seen at Las Animas, Colorado, in 1876, but he began to hear about better ones following the return home of Texans who had seen some top Herefords in the Central states and the East. Whereas he earlier had been partial to Shorthorns, Goodnight decided to strike a deal with the

*A portion of the famous JJ unit of the JA Ranch emphasized the quality of that early-day herd resulting from the consistent use of highclass purebred Hereford bulls. Note one of those bulls in the left foreground.*

persuasive O. H. Nelson, a cattle trader and bull salesman representing the firm of Finch, Lord & Nelson of Burlingame, Kansas, who had made a name for himself by shipping bulls by rail to Dodge City, Kansas, and then trailing them into the Panhandle. The Kansan, who came to be widely known as "Bull" Nelson, in 1883 sold to Goodnight and Adair their first Herefords.

It was a group which included not only 25 purebred bulls but also a substantial set of grade Hereford cows and calves which brought the Whitefaces to the JA range. Nelson arrived in the middle Panhandle region with them after six weeks on the trail. Even so, these Herefords which had originated in Kansas, Missouri and Iowa still impressed Goodnight to the extent that he paid $250 a head for the bulls, $75 each for 625 cows, an additional $75 a head for the approximately 400 calves following their dams, and $75 each for the dry cows and heifers.

Despite the fact that some of the other old-time ranchers expressed their suspicions that the well-to-do Adair was being "taken for a ride," J. Evetts Haley, in his Goodnight biography, pointed out that these cattle "proved to be the cheapest the ranch ever bought for they early established the reputation for quality of the JA cattle, the first in that region to sell for as high as twenty dollars as yearlings."

Those who had scoffed at Goodnight were impressed with the performance of the JA Herefords, with the result that during an eight-year period Nelson took more than 10,000 purebred bulls to the Panhandle, and is credited by those knowledgeable on the subject with having done more than any other one man to first establish the Hereford breed on the plains of the southwestern cattle country. Whereas Nelson had once brought some Shorthorns along with the Herefords, the demand turned so strongly for Herefords that ultimately he brought only Whitefaces. Nelson thus became a genuinely stalwart contributor to the advance of the Hereford breed.

A point of interest is that Goodnight put the best of the imported cows into a canyon pasture which he had fenced off with what is said to have been the first wire fencing in the Panhandle-Plains region. These, served by the best of the purebred Hereford bulls brought in, became the JJ herd, the primary purpose of which was to supply bulls for the great JA commercial herd. Thus, Goodnight pursued to the ultimate point the concept which had originated with him as a youth in the Palo Pinto country.

When the Goodnight and Adair partnership was terminated in 1888, the former became the owner of the Quitaque division of 40,000 acres, located on the plains north from the canyon. On it the little town of Goodnight was established, and there he built a commodious and handsome two-story home where he lived until his death at the age of 93 in 1929, continuing until near the end to breed and experiment with cattle, frequently with crosses involving buffalo.

Within sight of that home, which still stands, rests today the mortal remains of the man who did more than any other individual of his time to introduce and popularize the Hereford breed in the great Texas Panhandle cow country.

*Margaret (Mrs. Donald R.) Ornduff looks across the broad floor to a distant wall of Palo Duro Canyon, a factor which simplified the fencing problem when Colonel Goodnight first drove his herd to this location to commence a great chapter in ranching history.—Photo by the author.*

# KIRKLAND B. ARMOUR

KANSAS CITY, MISSOURI

1854-1901

IT was a great coup for the Hereford breed when Kirkland B. Armour of Kansas City, Missouri, decided to cast his lot with Herefords. Not only was he a member of a family famous throughout America for its success in the meat-packing industry—which seemed likely to give an aura of approval to the breed's claim to supremacy in the beef-making field—but this man who headed the Armour family's Kansas City operations was generally recognized as a great organizer and leader. These elements accrued greatly to the advantage of the breed.

While Kirk Armour's career as a Hereford breeder was relatively brief, it was meteoric in its rise. In fact, his beginning in the Hereford field was so auspicious as to attract industry-wide attention almost immediately. It happened this way: P. D. Armour, founder of the Armour packing dynasty, and uncle of Kirk Armour, was a friend of C. M. Culbertson, one of the early breed leaders and improvers in America. With the purebred business in the doldrums in the early 1890's, Culbertson's advancing years led to his decision to sell his herd. He thus requested P. D. Armour to ask that company's packinghouse buyers to mark up the Culbertson purebreds for all they would stand for slaughter.

The senior Armour responded that it was a shame to sell for beef that class of stock, and then he made a suggestion. He would inquire, he said, of the possible interest of his nephew, Kirk Armour, who had a farm near Kansas City, in buying the Culbertson cattle. If the younger man liked the idea, he continued, why not have the packinghouse buyers figure the herd for what it was worth over the scales, with Culbertson then to add whatever sum per head he thought was fair, with the cattle to be transferred to Kirkland B. Armour?

Both the younger Armour and Culbertson agreed to this proposal, and in 1891 the cattle arrived at the farm near Excelsior Springs, Missouri, which Kirk Armour owned.

He immediately took a liking to the Herefords and resolved very shortly to materially enlarge and strengthen the herd. Even though that first farm was not really distant from Kansas City, Kirk Armour decided also to find a farm nearer to his home in Kansas City so he could visit, observe and study the herd more frequently. He thus bought a one-section farm located some six miles south of his city home, and only a little farther from the packing plant which was situated near the junction of the Missouri and Kansas rivers. This area now, as a point of interest, is one of the choice residential districts of Kansas City, Missouri, and the Kirk Armour farm home still stands

*During a meteoric career in the Hereford field, Kirkland B. Armour became the most extensive importer of Herefords from the breed's English homeland in the history of American Herefords. These three heifers from the herd which Queen Victoria maintained on a royal farm near Windsor are representative of the quality he sought. Note Windsor Castle in the background in this sketch by George Ford Morris, a leading animal illustrator of the late 19th century, suggestive of the environment in which these heifers were raised.*–Breeder's Gazette *photo.*

*This picture, taken in 1980, shows the farm home where Kirk Armour entertained Hereford breeders and stockmen from near and far when they came to inspect his herd. It was on a property of over 600 acres, then located well south of the limits of Kansas City, Missouri, but long since surrounded by major residential districts of the city. The house was built in the latter 1890's and still stands at 6740 Pennsylvania Avenue.—Photo by the author.*

impressively in the midst of it. In the heyday of Armour's Hereford activity, few were the important Hereford personalities of the times who were not guests there at one time or another.

Possessed of ample means and bringing genuine personal enthusiasm to the farm and Hereford operations, he bought liberally of herd-improving stock both at home and abroad. Not only did the herd shortly become one of the largest in the Middle West but very soon it achieved a reputation enjoyed by few others in the history of the breed in America. And even fewer attained comparable stature in such a brief period of time.

He made two major importations from leading herds in the breed's English homeland, totaling 237 head, and a final shipment of 219 head had been arranged for arrival in 1901. Unfortunately, Kirk Armour did not live to see this shipment reach his farm because of his premature and untimely death on September 27, 1901, while the cattle were in quarantine at Baltimore. But his brother, Charles W. Armour, decided not only to carry on the Armour Hereford tradition, but in 1903 made an additional importation of 112 head from Herefordshire. It is a matter of interest that the two Armours brought across the Atlantic a total of 568 purebred Herefords, a number unmatched by the purchases of any other family in the history of the breed in America.

These purchases, plus the Culbertson Herefords and their descendants, provided a wonderful basis for the Armour herd. The first significant Armour herd-bull acquisition was Kansas Lad, a grandson of Anxiety 4th. Throughout his years as a Hereford breeder, Kirk Armour upon occasion sought the counsel of Frank Hastings, a Kansas farm boy who had become an employee of the packing company in 1889.

As time passed, they became great personal friends as well as employer and employee. They frequently inspected and studied the Armour herd together, and also became regular visitors to the Gudgell & Simpson farm, located in the same Missouri county, where they greatly admired the descendants of Anxiety 4th. On one of these visits, Hastings once related, he asked "Governor" Simpson about the importance of heart girth in a bull. "Simpson stood for a moment, looking the bull all over," Hastings later related, "and then said, 'Everything,' adding that the 'heart girth will carry him over hard times.' " Both visitors regarded this as a very significant statement.

This was an era in which rangemen were becoming more and more interested in the Whitefaces, and bulls bred by Kirk Armour helped greatly in building Hereford influence in commercial beef herds of the West and Southwest. The first crop of bull calves dropped in the Kirk Armour herd went to New Mexico buyers when around nine months of age at $45 a head. The second year's crop brought $65 a head to go to the great Prairie Cattle Co.'s Colorado range. Then a substantial lot went to the Matador range in the Texas Panhandle. After that the Armour bull output began to go in smaller lots to various range cattle growers, the prices increasing as the low times of the early to mid-1890's passed. So it is appropriate to emphasize the Armour influence in helping popularize the Hereford in the West by sending bulls of superior conformation and quality to the ranch country where they left the breed's indelible imprint upon range herds regardless of the caliber of the cows with which they were mated.

Every rancher knew the Armour name and all seemed to sense that the implicit Armour advocacy of this breed was sufficient evidence of the Hereford's beef-producing superiority. It was said of Kirk Armour that he was better known personally among ranchmen than any other man not actually engaged in the range industry. His firm conviction was that the best results on the range could come only from the use of registered bulls; if not Herefords, then the best of some other breed.

This personal conviction was expressed in the catalogue of a 1900 auction sale staged in Kansas City in conjunction with James A. Funkhouser, then a prominent and successful breeder of Plattsburg, Missouri: "The man who

*This was the plant of the Armour Packing Company, located immediately west of the Missouri-Kansas state line near the point at which the Kansas River joins the Missouri. Kirk Armour was president and general manager of the company. At its zenith, around the turn of the century, the plant contained 90 acres of floor space and employed 5,000 people. The buildings were razed in 1964 and 1965.—Illustration from the collection of Mrs. Sam Ray.*

sticks year in and year out to the use of a purebred bull," Armour wrote, "and continues to grade his cows up as high as possible, will make a better showing than a man who economizes on a bull from the fear that cattle will not always bring their present value."

Armour chose not to show his Herefords at the fairs, but that his output could have competed favorably with that of any of the exhibitors of the times is apparent in the fact that the sensational Prime Lad, which W. S. Van Natta & Son of Indiana exhibited to the senior and grand championship at the St. Louis World's Fair in 1904, was a product of the Armour program. His sire, bred by Armour, was a son of the Armour herd bull, Kansas Lad, while his dam was the English-bred cow, Primrose, a cow brought from England in one of Armour's importations.

This man's remarkable breed leadership was embraced whole-heartedly by the Hereford fraternity and Kirk Armour was elected president of the Hereford Association in 1898 after only a few years of activity in the field. His term coincided with the planning and staging of the first National Hereford Show, predecessor of the American Royal Livestock Show, which was held at Kansas City in the fall of 1899.

Incidentally, it was at this show that he first presented the handsome sterling silver Armour Cup to the exhibitor of the sweepstakes winner of the event's bull division. Frank A. Nave of Attica, Indiana, was the first recipient, on the basis of the victory of Dale in this contest. For the record, subsequent winners of the Armour Cup were Perfection, a son of Dale, shown by Thomas Clark of Beecher, Illinois, in 1900; Dandy Rex, in 1901, as a member of the show herd of Gudgell & Simpson of Independence, Missouri; and, in 1902 and 1903, March On 6th and Onward 4th, respectively, entries of J. A. Funkhouser of Plattsburg, Missouri. The latter's consecutive victories retired the cup from competition.

So successful in every sense was the initial Kansas City endeavor—to which Armour gave great personal leadership from both a community and industry viewpoint—that he was elected president of the breed association for a second term in 1899. His untimely death in 1901 at the age of only 47 cut short a career that in many ways seemed only to have just begun.

In further indication of Kirk Armour's civic spirit, it is appropriate to cite his contribution of Armour Rose, hailed as one of the best heifers ever produced in his herd, for the benefit of the building fund for a new Kansas City convention hall. A munificent sum for the fund was raised by the sale of raffle tickets on her, with the lucky woman who held the winning ticket then selling the heifer back to the donor for cash, after which Armour sold her in the auction held in conjunction with the show, with the proceeds to go to the building fund. She went to Gov. John Sparks of Reno, Nevada, for $2,500, a remarkable price for a heifer at that time.

So it is readily evident that Armour not only was a Hereford leader but also a civic-minded citizen whose influence, in a sense, still typifies not only his home community but also the cattle industry to which he contributed so notably.

# C. C. SLAUGHTER

DALLAS, TEXAS
1837-1919

No name was more famous in ranching circles during the heyday of his career than that of Col. C. C. Slaughter of Texas. Born in 1837, he was said to have been the first white male infant born there after the inception of the Texas Republic. His father was George Webb Slaughter, a native of Mississippi, who had served as General Sam Houston's right-hand man as Texans fought for their independence. The youngster, whose full name was Christopher Columbus Slaughter, began at an early age to assume manly responsibilities, and it is not too extravagant to state that from humble beginnings he pulled himself up by the bootstraps to build his fame and fortune.

At the age of 12, "Lum" Slaughter, as he came to be known in token of the second syllable of his middle name, participated in his first long cattle drive, and within the next half-dozen years became a full-fledged cattleman. According to the late Lewis Nordyke, West Texas ranching historian, the first cattle that went up the Chisholm trail from Texas to the railhead at Abilene, Kansas, were C. C. Slaughter cattle.

Born in Sabine County, in East Texas, he began early to help his father in farming, freighting and doing whatever else he could to add to the family's meager income. Eventually around 100 cattle were accumulated and one of the lad's duties was to assist his father in tending them. The father worked equally hard at farming and in preaching the gospel at pioneer settlements in East Texas. As the latter activity expanded, the son's responsibilities grew.

As more and more settlers moved into the Sabine country, the Slaughters decided to move on. The 15-year-old "Lum" drove one of the wagons when the family headed in 1852 for Freestone County, Texas, below present-day Corsicana. Within a few years, having heard that cattle-growing possibilities were good farther to the west, the father moved his young family to Palo Pinto County, west of Fort Worth.

There he continued the combination of ranching and preaching, interpreting the Word to listeners on the hard benches of the Baptist church, usually with a brace of pistols hanging from his belt and with a long-barrelled rifle within easy reach, in case the Indians who loitered thereabout should become troublesome. Busy as he was on Sundays as a parson, George Slaughter spent his week days with a lariat and branding iron, and doing other ranch work, until by 1871 he and his son had built up a herd of some 1,500 longhorns.

Even before the herd's sale to the Loving family in that year, the attention of both men had been drawn toward the possibility of making money by trailing Texas cattle northward to Kansas. They became partners in this enterprise, the trail herds sometimes including cattle which they owned and others belonging to other ranchers who contracted with the Slaughters to take them to the shipping points. It is recorded that they moved some 13,000 head north in the half-dozen years immediately preceding 1875, when the

*This was Col. C. C. Slaughter's celebrated "prairie pullman." Sometimes pulled by four matched mules, and at other times by six, according to the reports of the times, it was used by the owner to meet ranch visitors at the nearest rail point, Portales, New Mexico, and transport them across the 80 miles to the Slaughter properties where the principal registered Hereford units were maintained. Colonel Slaughter, wearing a long coat, stands beside the "pullman" here, with T. F. B. Sotham of Chillicothe, Missouri, next to front wheel.*—Breeder's Gazette *photo.*

*Colonel Slaughter turned the eyes of Texas, and the entire Hereford world, in his direction when he paid $5,000 for this bull, Sir Bredwell, in T. F. B. Sotham's sale in Kansas City in 1899. The champion at Omaha's celebrated Trans-Mississippi and International Exposition in 1898, the price was the highest ever paid for a bull by a Texas buyer to that date.—Cecil Palmer sketch.*

partnership was dissolved. And, mainly, it was found to be a profitable enterprise.

All the while, "Lum" Slaughter had been dreaming bigger dreams, and when the father's retirement was dictated by ill health, he was ready to surge ahead. Even though the frontier was wild and woolly, he was undaunted by the risk. He had served for a while in the Texas Rangers during the 1860's and knew how to handle himself. Most of his cattle projects, both in ranching and trailing, had returned a profit, and with his successes he began to expand his holdings, in both land and cattle.

He quickly gained recognition as a major factor in the Texas cattle industry, all the while also tending his civic duties. Following in his father's Baptist footsteps, he served for years as chairman of the Baptist General Convention of Texas. And he was one of the first to recognize the need for a cattlemen's association. With a few others, he helped organize at Graham, Texas, in 1877 what became the Texas and Southwestern Cattle Raisers' Association, of which he was the second president.

Always with his eyes toward the west, Slaughter's operations between 1877 and 1905 expanded into a ranching kingdom, located principally west of the Colorado River. On an expanse of 1,373,000 acres, some owned, some leased and some consisting of free range, he ran, according to various estimates, from 40,000 to 60,000 cattle. Although his holdings were not contiguous, they stretched from a few miles north of Big Spring to the vicinity of Morton, Texas, west of Lubbock.

Range country historians have credited Colonel Slaughter with having conducted at its peak, in the latter years of the 19th century, the largest individually-owned cattle operation America has ever known. His far-famed Long S brand was burned into the hides of as many as 12,000 to 15,000 calves annually. Every year over a substantial span of time it was said that more Long S cattle were marketed than of any other brand in the country. He reportedly became the biggest taxpayer in Texas and at the same time was a leading benefactor of the church, his community and other good causes.

His oft-stated philosophy as he advanced toward the pinnacle of the Texas cattle industry was expressed in his own words: "Keep all your heifers and raise all the cattle you can."

When a man of such stature became an all-out advocate of Herefords, as he did in the 1890's, the ranching world sat up and took notice, to the great advantage of the Whitefaces. Many others followed his lead, a fact which entitles him to lasting recognition in Hereford annals. While he had started with native Texas cattle, followed by Shorthorn crosses, it was the Hereford infusion on a grand scale which dotted his immense range with white faces and commanded the attention of the ranching world as the Slaughter cattle grazed as far afield as the Wyoming and Montana pastures which the Colonel sometimes leased in dry seasons.

In 1884, following the lead of the legendary Charles Goodnight, he bought 10 carloads of Hereford bulls and turned them out on the open range. In 1895 he bought from John Scharbauer of Midland, Texas, 1,900 Hereford females that had comprised the Cross J herd which was Goodnight's portion of the highly acclaimed JJ herd which had originated on the historic JA Ranch under Goodnight's direction. Of this set of Herefords, Goodnight acclaimed it "the best herd of cattle now in the round world." Sixty registered Hereford bulls came along in Slaughter's purchase from Scharbauer, who had been an extensive purchaser of Gudgell & Simpson breeding stock.

As his enthusiasm for Herefords grew, Slaughter made a trip in 1897 to the Central states of the U.S.A., during which he bought 57 additional Hereford bulls which reportedly cost him $50,000. He paid George S. Redhead of Des Moines, Iowa, $2,500 for the English-bred Ancient Briton which had won the sweepstakes award over all other Hereford bulls at the great World's Columbian Exposition at Chicago in 1893, in which contest he defeated Gudgell & Simpson's famous Don Carlos.

Colonel Slaughter, by now a Hereford enthusiast of the first rank, commissioned O. H. Nelson of Burlingame, Kansas, then the leading salesman of Hereford bulls in the West Texas range country, to select bulls for him at such celebrated auction sales as those held by T. F. B. Sotham at Chillicothe, Missouri, in the latter 1890's. The climax of these acquisitions came in the Sotham auction of March 1, 1899, when Colonel Slaughter was present in person to bid $5,000 and thus gain ownership of one of the most heralded bulls of his time, Sir Bredwell, which had been declared grand champion the previous year at the Trans-Mississippi and International Exposition at Omaha, Nebraska. This transaction set the bovine world agog because the price was the highest which had ever been paid to that date for an

*The hundreds of white faces evident in this picture make apparent the influence of the many high class Hereford bulls which was infused into the cattle on the Slaughter ranches. Wrote the reporter who was responsible for this picture taken just after 1900, "Herds of this sort are few and far between."*–Breeder's Gazette *photo.*

American-bred Hereford bull, not only by a Texan but by anyone else in the U.S.A.

There were other noteworthy Slaughter bull purchases as well, including a bull calf for which he bid $1,950 and a Sotham prospect which cost only $400 but matured into a bull valued easily at $5,000, rated on his merit as a sire. Some 60 others were bought from leading herds during this period, a group generally recognized, as once cited by Dean C. F. Curtiss of Iowa State College, "as the best lot of bulls ever put into range service." Nor did Colonel Slaughter consider this to be a gamble. He had already seen the result of the use of improved Hereford breeding in range production, having earlier scored a new record price of $7 per hundred on a thousand head of grass steers at the St. Louis market.

The most select of the Slaughter Herefords were placed on ranches he bought in Hockley and Cochran counties, west of Lubbock. These properties were referred to at first as the Sir Bredwell and Ancient Briton Divisions of the Long S. After 1901 these portions of the Slaughter holdings became known as the Lazy S. No commercial cattle rancher of his time gave greater impetus to Hereford prestige than did Colonel Slaughter in setting a highly successful example for his Texas and Southwestern neighbors.

With the acquisition of his highly valued Hereford bulls, Colonel Slaughter took another step which he deemed necessary in order to get full value from their service in his herds. He undertook an extensive fencing program which enabled him to fully utilize controlled breeding in his own herds and also to keep away from his bulls the cows belonging to less progressive ranchers. In this he also showed the way to others.

C. C. Slaughter's business activities were by no means confined to cattle and ranching. As early as 1884 he became a Dallas banker, obtaining in that year a charter for the American National Bank there. He built an office building in that Texas metropolis. But his personal interest was always centered on the range.

On a July day in 1976, this writer stopped briefly at a point in Cochran County, near Morton, Texas, which was designated as "C. C. Slaughter Ranch Headquarters," and once known as the Sir Bredwell Division. Still in place were adobe structures forming a quadrangle, apparently built to provide living quarters for ranch employees. Nearby was a small barn. In all directions stretched what once had been a vast expanse of range grassland, now under irrigation and entirely in crop production, mostly cotton. The glory of the old days had faded.

Following Colonel Slaughter's death in 1919, the process of breaking up the Slaughter holdings into farming entities began, and by 1957 it was completed. The final segment to meet this fate was a 23,300-acre unit in the northern part of Martin County, south of Lamesa, Texas.

There is one postscript. Had not fate intervened, the Slaughter influence might have continued well toward the end of the 20th century in the registered Hereford field. A grandson of the Colonel, John Henry Dean, Jr., during the 1930's, developed a highly successful herd located first at Lamesa, and later at Fort Worth. Tragedy struck when Dean, then 37 years old, died in an airplane crash in 1942 while flying a Civil Air Patrol mission during World War II.

*This barn still stands on what once was the Sir Bredwell Division of Colonel Slaughter's ranches. It is two or three miles southwest of Morton, Texas, in a region now entirely under cultivation.—Photo by the author.*

# W. H. Curtice

EMINENCE, KENTUCKY

1853-1924

Most ardent of the Kentucky stockmen who were converted to the Hereford cause by the celebrated 1886 "invasion" of the Bluegrass state by show herds of Whitefaces representing several leading northern breeding establishments was Col. W. H. Curtice of Eminence. What he saw when these strings competed at Shelbyville and Lexington, in the heart of a region which theretofore had been dominated by Shorthorns, led him from that time forward to become one of the Hereford breed's most ardent disciples.

His first purchase was of a Hereford bull, which was used on 20 grade heifers. Fourteen of the resulting calves were fed out as steers, and when sold on the New York market brought a price well above that paid for any other export steers that year. He was convinced.

Curtice first established a reputation in Hereford circles through the use of Lord Wilton and Garfield bloodlines, mainly through the service in his herd of Earl of Shadeland 9th, bred of Adams Earl of Indiana, and acquired along with two foundation cows from Pickett & Bailey of Finchville, Kentucky. But his lasting fame rests upon his purchase a few years later of the Gudgell & Simpson-bred Beau Donald. And his still later acquisition of Perfection, of Garfield and Anxiety descent, which as a young bull had been a remarkably successful showring winner for Thomas Clark of Illinois.

But it was Beau Donald which put Colonel Curtice on the Hereford map, so to speak. The element of chance never was better illustrated, perhaps, than in this bull's story. Oblivion was but a step away when Dame Fortune interceded. Depressed by a general economic downturn that came near to reaching panic proportions, the market for purebred cattle dried up during the early 1890's.

One example will serve to highlight how serious this shortage of buyers became, even for such a heralded herd as that of Gudgell & Simpson. Their bull demand was so poor in 1893 that the entire crop was castrated with a single exception, and that individual barely escaped. "Governor" Simpson, partner in the firm and famous for his good judgment, had vowed that if he couldn't get $100 a head for these calves he would make steers of the entire lot. Only the timely appearance of H. B. Watts of nearby Fayette, Missouri, saved the youngster which brought lasting fame both to himself and to Colonel Curtice. The price Judge Watts paid for this intensely Anxiety 4th-bred bull–sired by Beau Brummel by Don Carlos by Anxiety 4th and out of Donna by Anxiety 4th–founder of a family whose descendants continued to be held in high regard well into the middle years of the 20th century was $100, as Simpson had vowed.

At just the right time, when Curtice was seeking a new herd bull, Beau Donald had, through such sons as Lord Erling and Prince Rupert, among others, distinguished himself in the Watts herd as a sire of show cattle and top breeding prospects. Among those lending a hand in the Kentuckian's search was Secretary C. R. Thomas of the Hereford Association, through whose assistance Curtice went to the Watts farm to inspect the bull which Watts had agreed to sell if he could get a price of $1,000. This was in the summer of 1897 when Beau Donald was four years old.

Curtice was impressed but tried not to show it, exclaiming, after a studied examination: "Turn him out; I never saw a bull I'd pay $1,000 for."

Watts calmly replied: "Very well, I'm glad to hear you say that; and rather than sell you 'Beau' I'll give you a check for $100 to release me from my proposition." The Kentuckian was noncommittal.

The Missouri man accompanied his visitor to the railroad station for the beginning of the latter's trip home. As the whistle of the approaching train sounded, Curtice broke the silence: "Ship 'Beau' to me in time for him to reach the fair at Shelbyville on August 24."

This proved to be a historic moment, and the ensuing years added glowing chapters to breed history through Beau Donald's breeding performance at Pine Park Place, as the Curtice farm was known. According to contemporary breed historians, he transmitted with such remarkable consistency to his get

*Col. W. H. Curtice, typifying the traditional Kentucky colonel, astride his favorite mount, the beautiful Champagne, ready for a ride into a Hereford pasture.—Photo by courtesy of the late Curtice Martin.*

his marvelously straight lines, strong front and superb quarters, together with his elegant style, finish and symmetry, that the Curtice show herds moved straight to the front row of the Hereford parade. And Watts accurately remarked later on to "Beau's" new owner, "I am the architect of your fortune."

Mostly from dams of Earl of Shadeland 9th descent, the get of the 2,400-pound Beau Donald won remarkably at the major shows, accounting for, among other victories, the blue ribbons for get of sire at the American Royal shows of 1904 and 1905, and at Chicago's International they made it three in a row in that class—1903, 1904 and 1905.

Fellow Hereford breeders had been visibly impressed with Beau Donald's get even earlier. Youngsters sired by him were first offered at public auction in an association sale at Kansas City, where 16 of them commanded an average of $629. But overall, the Colonel perhaps was proudest of his show herd's performance at the St. Louis World's Fair in 1904, since it accentuated the uniform excellence of his output. The following words were written in 1957 by his grandson, Curtice H. Martin of Montana, concerning the outcome of that competition:

"With some 26 U. S. exhibitors and one from England, it was called the greatest Hereford show to that date. Again the Colonel and his greatest of herdsmen, Jim Hendry, had out a full herd sired by Beau Donald 58996. The rejoicing was great in the family the day when Beau Donald's get won first and fourth for the Colonel, and Overton Harris, his great friend and toughest competitor, from Harris, Missouri, ranked fifth on the get of Beau Donald 5th, a son of 'Beau,' while Gudgell & Simpson were seventh on the get of Beau Brummel, the sire of Beau Donald. Also, the Colonel won first and third on produce of cow, all of these, of course, calves sired by Beau Donald."

Feeling that it was impossible to get too much of a good thing, and determined to even more strongly fix the old bull's desirable characteristics, Colonel Curtice later purchased from Gudgell & Simpson a full-brother to Beau Donald, seven years younger, and strangely enough, also named Beau Donald. They were distinguished in a pedigree only by their own registration numbers. Crossing "young Beau," as he was called, on the daughters of "old Beau," proved to be a master stroke in getting some of the herd's best-producing females, according to Curtice Martin, who worked as his grandfather's right-hand man for several years.

Scores of breeders acquired sons of old Beau Donald during the earlier years of the 20th century, and their influence carried on far beyond. Beau Donald 5th was a key foundation bull and a great sire of winners for Overton Harris. George Chandler of Baker, Oregon, used a son of old "Beau" in that still-existing herd's earlier period.

Luce & Moxley of Shelbyville, Kentucky, founded a noted herd with Prince Rupert by Beau Donald and a set of Beau Donald females. A notable double-bred Beau Donald sire was Good Donald, founder of a line of his own for Wallace and E. G. Good of Grandview, Missouri. Sons of the latter were used for years in the Tequesquite Ranch herd of T. E. Mitchell & Son, Albert, New Mexico, where descendants doubtless carry on, although by now many generations removed. At its peak, the Beau Donald influence was widespread in the industry.

For several years Colonel Curtice had been interested in and impressed by the Perfection family. This bull himself had been a remarkable winner in his younger years, including grand championships at both the American Royal and International shows at the turn of the century, and had been sold at auction in 1902 for $9,000. But it was the breeding performance of Perfection's get, including Perfection Fairfax, the herd bull of Warren T. McCray of Kentland, Indiana, and Defender, an American Royal champion for C. G. Comstock of Albany, Missouri, among numerous others, which led to Perfection's ultimate acquisition by the Colonel.

To acquire the old bull even then, when he was approaching 10 years of age, the Kentuckian had to buy the entire herd of G. H. Hoxie of Thornton, Illinois, which Perfection then headed. Even though, as Curtice Martin once related, only about two seasons of service were gotten from Perfection at Pine Park, the purchase added another feather to the Colonel's hat. In the words of Curtice Martin, "He nicked so wonderfully on Belle Donald cows (as the females by Beau Donald were named) that many outstanding herd sires were produced from this cross."

Best known of these was Beau Perfection 24th, sold to Col. E. H. Taylor, Jr., of Versailles, Kentucky, in 1914 at a then record price of $12,400, and soon afterward re-named Woodford. In later years, the Colonel often said, "He was the cheapest bull I ever sold."

Be that as it may, Woodford, along with 20 Beau Donald cows bought from Curtice at the same time, became the founder of a family of his own which was rated as one of the breed's most successful during the middle years of the first half of the 20th century. And the record might have been even more outstanding had not Woodford been destroyed in a fire at the Taylor farm after only four years of service there. Further, his descendants and influence were widely scattered following Colonel Taylor's death in 1924.

Other Beau Perfections bred at Curtice's farm also contributed to their

*W. H. Curtice's grand champion beef herd at the 1902 Ohio State Fair. This interbreed class, in which an Angus quintet ranked second and a group of Shorthorns third, enhanced the prestige of the Hereford breed in the eyes of the beef-cattle industry. The Curtice string was comprised entirely of the get of Beau Donald.*—Breeder's Gazette *photo.*

family's fame through their service in such herds as those of John E. Painter & Sons of Roggen, Colorado; A. B. Cook of Townsend, Montana; B. N. Aycock of Midland, Texas; Henry Thiessen of Culdesac, Idaho, and numerous others. Further, J. N. Camden of Versailles, Kentucky, founded his famed Hartland Farms' herd with 20 Beau Donald-bred heifers and a grandson of Perfection; and the Perfection influence, along with that of the Beau Donalds, "built" the herd of Col. W. H. Roe of Shelbyville, Kentucky.

When the value of land in the vicinity of Pine Park Place reached a high level, a decision was made to sell the property and move the herd to Canada. Thus the herd was transported to the Peaceful Valley Ranch at Shepard, Alberta, not far from Calgary, where it was situated at the time of Colonel Curtice's death in 1924. The Curtice show herds performed sensationally in western Canada and the U. S. Northwest, with the 2,600-pound Beau Donald 192d, strongly bred to Beau Donald and from a Perfection daughter, being undefeated in class in 17 consecutive shows, and becoming either grand champion or reserve champion at all of them. Beau Perfection 48th, double-bred to both Beau Donald and Perfection, also scaling 2,600 pounds, likewise won so heavily that he was reported to have "won as many championship ribbons as any other bull in America."

Belle Curtice Wright, the Colonel's daughter, inherited the herd from his estate and the cattle were moved to Stevensville, Montana, in 1927. Her son, Curtice H. Martin, bought the herd from her in 1940, continuing its operation until his death. Since then Martin descendants have carried forward the family Hereford tradition and ideals.

# Robert H. Hazlett

EL DORADO, KANSAS

1847-1936

Virtually every honor that can come to a man engaged in constructive livestock improvement was accorded Robert H. Hazlett of El Dorado, Kansas, during his long and productive career as a breeder of registered Herefords.

Probably no producer of purebred seedstock of any breed ever was more widely or frequently hailed as a master breeder. It was written that his name should surely be included in any livestock industry hall of fame, as it was long ago in the famed Portrait Gallery of the Saddle and Sirloin Club which gave world-wide recognition to the immortals in the field of animal agriculture.

Hereford breeders of America, on the occasion of his 80th birthday, gave him a gold medal inscribed: "To Robert H. Hazlett, America's Premier Hereford Breeder." He was the guest of honor at the 61st annual meeting and dinner of the Kansas State Board of Agriculture in 1932 when he was lauded as the "Premier Hereford Breeder of the United States." In that same year Kansas State University conferred on him the honorary degree of Doctor of Agriculture in recognition of his contribution to the study and development of the science of genetics and animal breeding.

"He had a genuine, heart-whole desire," said Henry J. Allen, former governor of Kansas and later United States senator, "to accomplish something really worthwhile, to live a life that would be remembered for what it had done. . . ."

Added William Allen White, famous editor of the *Emporia Gazette* at Emporia, Kansas, concerning his famous neighbor: "There was always a vision in his heart. . . . It was a general desire and his personal ambition to aid in effecting the betterment of his community, of his state, and of his fellowmen; to make life worthwhile by making the world richer and better for his having lived in it."

Twice the Hereford breeders of America honored him by electing him to the presidency of the American Hereford Association. The first time was 1908 when he was 61 years old and the next in 1932 when he had reached the age of 85. How he felt about the breed was expressed on the latter occasion in these words: "No man with an ambition to breed better Herefords, and who delights in their growth and development, can fail to be benefitted, both physically and spiritually, through his association with Herefords."

In 1936, when he was in his 90th year, members of the Association at their annual meeting in Kansas City unanimously elected him president emeritus, an office created especially in honor of this dean of American Hereford breeders, and a post never occupied by any other breeder.

All of these many tributes were in recognition of Robert Hazlett's unique accomplishments in the Hereford field. At the outset, it should be pointed out that he went all the way "on his own." He was born in a log cabin on his father's Illinois homestead. As a lad, he herded the family cows on the open prairie, and worked at other tasks commonly assigned to farm boys of those times.

He attended country school and then went to the just-opened Illinois Industrial University, now known as the University of Illinois at Urbana, from which he was graduated in 1870. After a year or two as a country school teacher, he entered the law school of the University of Michigan at Ann Arbor, receiving his diploma in 1873. He then began to practice law in Springfield, some 30 miles north and west of his birthplace.

His law practice prospered and he made money, just as he had when as a boy he had raised potatoes for sale, had worked at odd jobs, traded, bought and sold, and taught school to pay for the education which he was bound to have. He was elected and re-elected to various posts, and friends told him he could be elected to any office he might choose. It was a temptation, he later said, but he had a desire to get away from the confinement of a law office and the demands of politics. Seemingly a natural born money-maker, he had accumulated some savings which had been invested in Nebraska and Kansas land. He and his bride of a year moved to El Dorado in 1885.

He now retired from law and spent full time in real estate activity. He also had made an investment in a gold mine at Leadville, Colorado, and when problems arose there he spent some months giving the operation his personal direction. This activity proved so successful that a highly profitable sale of his interests was made, with the resultant profit being invested in more Kansas land, on which oil was shortly discovered. Oil and banking thereafter were his chief city business interests.

But he never forgot his rural beginnings and went often to his farm, just north of El Dorado, where he had started a grade herd of beef cattle. And he remembered having seen, as a boy in Illinois, some cattle of a then new breed–Herefords–at the state fair in Springfield. Older breeds were so entrenched that the Illinois State Board of Agriculture refused to make a classification for the Herefords. The young man believed that the newcomers were not getting fair treatment, and his sympathies were with them in their fight, and apparently ever afterward.

When the opportunity arose a dozen years or so after the Hazletts moved to El Dorado to buy the nearby herd of 16 Herefords belonging to H. H. Grover he made the deal. In his later years he said that he had gone often to look at Grover's little herd, and that their general appearance and red and white markings appealed to him, strengthening the sympathetic interest he already had in this breed's favor. He quickly added that he knew very little about cattle of any breed. But he did know that the group which comprised his first

*During a field day at Robert H. Hazlett's Hazford Place in 1927, these three stopped long enough for the photographer to picture them. Hazlett is in the middle, with R. J. Kinzer, secretary of the American Hereford Cattle Breeders' Association, at left, and Will Condell, long-time superintendent at Hazford Place, at the right.—Smith & Morton photo.*

Hereford purchase in 1898 included 14 cows and heifers and two bull calves, and that all were said to be strong in the blood of a bull named Anxiety 4th.

Recognizing his own limitations, he was first influenced by what many long-time breeders advised: That his cattle were so closely linebred that it was necessary to find an outcross. With that in mind, he bought three outcross bulls during the next few years. All were good-looking individuals and were of popular bloodlines–and all were disappointing. From all of the heifers sired by those three bulls, only one was kept in the herd.

In the meantime, Hazlett recounted many years later, he did some investigating and thinking on the science of breeding. What he learned confirmed his own early experience that the results obtained from the closebreeding, or linebreeding, which he was following, had been so much more satisfactory in yielding a higher percentage of good calves than the outcross bulls had produced that he made up his mind to go back to the Anxiety bloodlines and stay with them unless he learned from experience that the closebreeding would destroy the usefulness of his herd.

After that, he used four sons of the Gudgell & Simpson-bred Beau Brummel, a 2,600-pound grandson of Anxiety 4th that had been a successful show bull, with results which made breed history. Subsequently he "rescued" the Gudgell & Simpson-bred Publican, of similar ancestry, from the Texas Panhandle range of the Matador Land & Cattle Company, and proceeded to infuse that bull's influence constructively into his Hazford Place herd. In the immediately following years as Hazlett herd bulls came various descendants of the original bulls, along with a few others of similar bloodlines.

Of slightly alien ancestry was Hazford Rupert, a bull that made an outstanding contribution to the Hazlett herd and to the breed in general. Purchased in dam in 1916, he was a great-grandson of Prince Rupert, noted son of W. H. Curtice's Beau Donald that, in turn, was a famed son of Beau Brummel and out of a daughter of Anxiety 4th. The Hazford Rupert blood, especially through a son, Hazford Rupert 25th, whose dam was a granddaughter of Publican, nicked splendidly with the blood of the Beau Brummel descendants, particularly the Bocaldo line which traced through Printer to Beau Brummel, to yield the combination which became generally known as "Hazlett breeding."

From that point onward no outside bulls ever served to any significant extent at Hazford Place–a farm name comprising the first part of the Hazlett name and the last half of "Hereford." And it was the same with the cows comprising the breeding herd.

In addition to the linebreeding program practiced so effectively in his herd, Hazlett also was a stickler in the matter of culling out such animals as did not meet his high standards. Nor did he sell such animals to other breeders. For years he was occasionally criticized because he chose not to sell females to other purebred breeders. If they were good enough to go into production, it was his view that they should go into the Hazford Place herd to replace older cows to be sent to market; otherwise, they should go to the beef packer or, possibly on occasion, into commercial production without the transfer of registration certificates. Likewise, he insisted on the castration of quite a large percent of each year's bull crop; in fact, as he once stated, he thus disposed of "all those that I did not believe would give good results as sires, wherever they might go, whether as range bulls or herd bulls."

He once expressed his philosophy along this line in these words: "The matter of selection is at least as important as bloodlines. For many years I have disposed of cows and heifers, in the yards for beef, because of some defects or weaknesses in conformation, quality or character that I believed should not be perpetuated. To say that no animal of any line of breeding, of either sex, having any serious fault or weakness should be used for breeding purposes in a purebred herd is to state an elemental or fundamental fact."

Strict adherence to this policy throughout his almost 40 years as a breeder, after waiting until the age of 51 to buy his first registered Herefords, resulted in a herd which became world famed for its uniform excellence. Show records confirm this point. Hazlett entries at the American Royal placed as high as second a few times during the period from 1909 to 1914. Then in 1915 three of his entries captured blue ribbons. The next year one of those three, Bocaldo 6th, came back to the Kansas City show, and at all others in which he competed during that show season, to win the senior and grand championship award. He, of course, was strictly of Hazlett breeding.

From that point until the very last show in which Hazlett Herefords competed, they reflected enormous credit upon the man who bred and exhibited them. That last show was the 1936 Chicago International where, from the wheelchair to which he was confined because of failing strength, he watched as his Hazford Rupert 81st and Bonita Zato, products of his breeding program, were adjudged the champion bull and champion female of that great event. What a fitting climax! In less than a month he died, full of years and full of honors.

Probably no exact record exists of the total number of ribbons won by Hazlett Herefords. A partial list, however, is spectacularly impressive. During the 20 years between 1915 and 1934, the herd captured no fewer than 649 blue ribbons, 401 blues and 189 championship purples. Hazlett entries continued to win heavily through the 1935 and 1936 campaigns, and undoubtedly accounted for a few dozen more similarly high awards. And, bear in mind,

*Robert H. Hazlett with two of the scores of champions produced at his Hazford Place during the herd's nearly 40 years of existence. The bull here is Zato Rupert and the heifer Iza Rupert. Full-brother and -sister, sired by Hazford Rupert 25th and out of Izatone by Hazford Tone, they were pictured following championship victories at the 1933 American Royal, a feat which they duplicated shortly afterward at the Chicago International.—Smith photo.*

that all of these winnings were accomplished in the country's leading Hereford expositions where the awards were most hotly contested.

When asked, on the occasion of a field day at Hazford Place in 1925 to tell something of the methods by which he had built his herd, this exceedingly modest man said, in part:

"A good start doesn't make a good herd. You must keep everlastingly at it. You have paid me the compliment of saying that I have accomplished a great deal with this herd. I have not, by any means, reached perfection. I am satisfied that further improvement can be made, and I am going right ahead along the same line in the faith that I shall be able to make it.

"I have been asked to say what I consider the principal factors in the building of a good herd of Herefords. I would answer that they are energy, patience and perseverence, not too much theory, and plenty of common

*No one ever before had seen a crowd at a Hereford sale as large as that which attended the Hazford Place dispersion which began on June 15, 1937. This gigantic circus tent, with seats for 3,000 persons, was filled to over-flowing, with the aisles and approaches also packed shoulder to shoulder. When the sale ended on the third day, the herd had been distributed to buyers from 26 states and Canada. The heifers pictured here awaited their turn in the ring.*

*Biding their time as the Hazlett dispersal progressed, this set of heifers found relaxation in a pasture near the Hazford Place headquarters buildings.*

*This famous pair of Hazford Place females grazed contentedly in a nearby paddock after going through the salering. At right was Bonita Zato, undefeated grand champion female of the 1936 season, which was sold for $3,100, a new seven-year-high, while at left was the 10-year-old Iza, then the highest-ranking female in the Register of Merit, for which the last bid was $2,050. Both went to Long Meadow Ranch of Prescott, Arizona.—Photos by the author.*

sense. There will be disappointments. The results will not always be what you expect them to be. The science of breeding is nothing more than the application of the facts, as we know them, about the production of animal life.

"There is much in heredity. Animals run true to species, in conformation, characteristics and quality. Like begets like, or the likeness of some ancestor. I believe you will get a larger percentage of good animals by that kind of mating. Of course, there occasionally will be a disappointment. . . .

"The matter of selection is very important. Of course, a good herd bull is an essential. Every breeder tries to get one, but they are scarce. There cannot be a good herd without good females. During my first few years I did not realize the necessity of culling the heifers. But after a few years I made it a rule to cut out all common or plain heifers, and fatten and sell them for beef. I believed a heifer that wasn't good enough for my herd was not good enough for the herd of anyone else.

"Any normal young man can do what I have done here, and more," the veteran breeder said in conclusion, "if he sticks to it long enough, and if he also has an inherent love of animal life–especially young animal life."

Some six months following the owner's death in December, 1936, what was generally acclaimed the world's greatest Hereford herd of its era was dispersed in one of the most spectacular beef cattle auctions of all time. Newspaper accounts estimated the attendance on the first day of the three-day sale at 8,000. In a year when all registered Herefords averaged under $200, the 604 head averaged $505, setting a new record considering the number of cattle sold. It was later said that this figure might well have been eight or ten times higher had the sale been held a dozen years or so later when general price levels had risen spectacularly. The cattle went to 133 buyers, representing 26 states and Canada.

It was the culmination of this unique man's lifework with Herefords–a sale in which at the time it was pointed out that at least 99 percent of the cattle listed, along with their ancestors to the fifth and sixth generations, were bred and raised on his farm and under his direction.

GEORGE CHANDLER

HERBERT CHANDLER

# GEORGE CHANDLER

BAKER, OREGON
1845-1934

# HERBERT CHANDLER

BAKER, OREGON
1884-1973

STRANGELY enough, it was a set of calves from a bunch of Shorthorn cows that made a Hereford convert of George Chandler, who first put Baker, Oregon, on the Hereford map. It happened this way. James A. Funkhouser of Plattsburg, Missouri, in 1896 shipped a carload of grade Shorthorn cows to Oregon. Most of them had been bred to Hesiod 2d, member of a purebred Hereford family that was to contribute greatly to the national prominence of the Funkhouser herd. In George Chandler's own words, written in 1925, the result was as follows:

"When these cows dropped their crossbred calves it created a sensation among our cattlemen. An improvement on the Shorthorn? Impossible! At least, we had always thought it impossible! But there were the calves. They spoke for themselves. And there was nothing for us to do but say the impossible had come to pass."

As a native of Missouri, George Chandler had "to be shown," but when this demonstration of the potency of the Hereford influence "showed" him, he swung into action. He became a registered Hereford breeder in 1898, and succeeding generations of Chandlers have ably followed in his footsteps, including his son Herbert, his grandson Charles, and the latter's sons and their children.

The Chandler family's arrival in the Baker Valley of northeastern Oregon–then yet largely Indian country–was not exactly accidental, but neither was it in the original plans. George Chandler's father, possessing a spirit of adventure, and willing to face the challenges involved, had gone to California's Sacramento Valley in 1850, probably hoping to "strike it rich." However, the gold eluded him and he returned to Missouri in 1853. But having successfully made the round trip, his neighbors and others considered him a veteran of the western trails and overland travel.

So in 1862 the elder Chandler was easily persuaded to become captain of a California-bound wagon train. But this time he took along his family, including a son, George, then 17 years old, who piloted the ox teams hitched to one of the covered wagons. Along the way the train was met by stories of gold in Oregon. The magic word sank so deeply into the minds of the party that the train changed course from California to Oregon, and there it was that the elder Chandler and his family decided to stay–in a spot where there was not a house within a hundred miles. While gold was the drawing card, just as it was for so many pioneers westward, the Chandlers also changed from searching for the precious metal to the more certain returns from guiding ox teams hauling freight to gold mines in eastern Oregon and Idaho, which young George did for his first five years. Then he found just as adventurous a life in raising beef cattle on the range.

The Chandler name has been one to conjure with in any consideration of the course of the Hereford industry in the West in all of the years since 1900. George Chandler's 1898 purchase of registered Herefords consisted of a bull and a heifer acquired from J. W. McKinney of Turner, Oregon. He had been impressed by a picture of St. Elmo of Shadeland and four of that bull's get which he had seen in an advertisement of Z. T. Kinsell, pioneer Hereford breeder of Mt. Ayr, Iowa, in *The Breeder's Gazette*. So he journeyed, also in 1898, to Iowa and bought nine daughters of that noteworthy herd-header which was a son of the famous Garfield and out of a dam by the equally famed Lord Wilton, which headed the great English herd of T. J. Carwardine, in its later years. At about the same time he got seven daughters of the Gudgell & Simpson-bred Princeps, the herd bull of Steele Bros., in Kansas.

Having embarked on this course, George Chandler was aware that he would soon need a herd-header since he had decided not to use the

McKinney bull. In 1898 there were few names which loomed more brightly on the Hereford horizon that that of T. F. B. Sotham, whose Weavergrace Herefords were located at Chillicothe, Missouri. Lured perhaps also by the fact that this was right back close to the family's old Missouri home, he went to the Sotham sale, carefully studied the bulls listed and marked his catalogue as he awaited the opening auction gavel of the veteran Col. M. A. Judy. He indicated a maximum bid of $500 for his choice, which was the yearling bull Excellent, a double grandson of the noted Sotham herd-header, Corrector. He was Lot 1 in the catalogue and was the first one sold.

The auctioneer opened the sale with a rousing speech and the Chandler choice was led into the ring. When Colonel Judy said, "Who will start him at $1,000," as Chandler later recounted, "some four or five hands went up," and, he added, "I think I shrank about four inches." Although, as the Oregon man later commented, he was knocked over the ropes in the first round, he at least had the satisfaction of finding that he knew how to pick a good one. Later in that same sale, at a much lower price, he bought a yearling son of Corrector.

After the sale was over he sought out the dam of his purchase and when he found her he was "knocked out worse than ever," as he once put it. She was the kind which he felt should have gone to the butcher, and this experience taught him a lesson he never forgot regarding the importance of conformation and quality in an animal's ancestors, which he felt were as essential as

*Herbert Chandler, left front, as president of the American Hereford Association, sat beside President Dwight D. Eisenhower amidst other dignitaries at the dedication of the new headquarters building of the Association in Kansas City on October 16, 1953.—Harkins photo.*

pedigree. Years later he said, "I wouldn't think now of using a bull, even if he were pure gold, from a cow like that."

A few years later he attended the International show at Chicago, where he met Clem Graves, a prominent breeder from Bunker Hill, Indiana. At the show's conclusion, he went home with Graves and purchased two bulls, one of them, Crusader, in which the influence of Earl & Stuart and Fowler & Van Natta was combined, and the other a son of Crusader. Despite the fact that Crusader was a $10,000 bull, Mr. Chandler did not find his get satisfactory, as he put it, and sold all of the females by him but one. So it is evident that performance, along with individuality and pedigree, all were considered important criteria by him.

A few years later, again needing a new herd bull or two, he wrote asking if Graves would buy for him what he considered the best two bulls available in Indiana, setting no price limit. One of the two selected was Beau Donald 33d, which became a key Chandler sire, and considered to be one of the three best bulls the Oregonian ever owned. Bred by W. H. Curtice of Eminence, Kentucky, he was sired by the great Curtice herd-header, Beau Donald, while his dam was strong in Earl of Shadeland ancestry. "The 33d" was considered by some to be the best producer in the Beau Donald family. Some 35 of "the 33d's" best daughters were retained for years in the Chandler breeding herd. His sons and daughters spread the bull's reputation across the Northwest.

The second of George Chandler's "three best bulls" was Debonair 26th, bred by C. A. Stannard of Emporia, Kansas. This bull, intensely bred to Gudgell & Simpson's Beau Brummel, crossed remarkably well on the Beau Donald 33d cows, and Chandler always felt that some of the credit for the bull's performance as a sire should go to "the 33d" as the sire of the cows involved. Eleven of the 15 Debonair 26th sons which comprised the Chandler 1917 Denver show's grand champion carload of bulls were from Beau Donald 33d daughters. Described as a very long-bodied, deep-fleshed, masculine bull, with superb hindquarters, Debonair 26th weighed around 2,100 to 2,200 pounds without benefit of any feed other than gleaned on pasture.

The third of George Chandler's "top three" was Prince Rupert 60th, bred by Luce & Moxley of Shelbyville, Kentucky, and double-bred to Curtice's Beau Donald. Not much to look at, according to Chandler, but his dam was a genuinely great cow, confirming the Oregon man's insistence on this point. The calves by Prince Rupert 60th made an immediate imprint on the herd, as indicated by the fact that the first crop contributed strongly to the Chandler show string, and his very first calf was sold at the Denver show for $2,150, a long price for the times.

Though not listed among his top three, another contributor to the herd

*In a herd founded more than 80 years ago, the champion Mark Donald presided here among a few representatives of Chandler Herefords at a spot where the pasture was abundant and water plentiful.*—Record Stockman *photo.*

foundation was the Gudgell & Simpson-bred Beau Mystic, a grandson of Beau Brummel. He was bought from Stannard.

Thus was laid the foundation for Chandler Herefords. On the job during the herd's developing years, closely observing and aiding his father in every possible manner, was his son, Herbert Chandler. He "developed" with the herd, in fact, and was ready by 1919 to take complete charge of the cattle, then numbering some 250 head. Carrying forward largely with the breeding stock already in the herd, plus their descendants, he guided the invasion by Chandler Herefords into most of the country's major shows where they acquitted themselves with remarkable success. But of far more lasting significance from the standpoint of history was their contribution to countless herds throughout the Pacific Northwest and Rocky Mountain West, plus other individuals which went still farther afield to spread the Chandler influence.

Then came the bull which marked a major turning point in the Chandler fortunes, and stamped Herbert Chandler's imprint on the herd which his father had moved steadily upward through the first 20 years of the 20th century. This was Prince Domino 164th, which was selected from the second-prize load of calves exhibited by Fulscher & Kepler of Holyoke, Colorado, in the carlot bull division at 1925's National Western Stock Show in Denver. This son of the history-making Prince Domino, and from a dam tracing to such breed-foundation sires as The Grove 3d, Majestic, Anxiety and Sir Richard 2d, contributed immeasurably to much that was later accomplished at the Oregon establishment and in the herds of its customers. It is of interest, in indicating his own individual merit as a second cornerstone of his contribution, that under Herbert Chandler's tutelage Prince Domino 164th returned to Denver in 1926 to win second as an individual in the show's junior yearling bull class.

In the Chandler breeding herd he proved to be a remarkable catalyst. It was a mating of this new bull with a cow in which was combined the blood of the four major George Chandler herd-foundation sires—Beau Donald 33d, Debonair 26th, Beau Mystic and Prince Rupert 60th—that marked a major milestone in the Chandler program. The result of that mating was Donald Domino 2041990.

Then came the next big step forward. When the noted Hartland Farms herd of J. N. Camden of Versailles, Kentucky, was dispersed in 1928, Herbert Chandler purchased the three-year-old Belmont Hartland. This bull, a grandson of Mischief Mixer by Mousels' Beau Mischief, and out of the W. H. Curtice-bred Belle Woodford 6th, representing the Woodford-Beau Donald cross, was one of the most heralded winners of his time. The price paid for him was $3,300, in a day when only a few bulls were selling for as much as $1,000. John Letham, then a *Hereford Journal* field representative, executed the Chandler buying order.

Always a stickler for conformation and quality as well as sound inheritance, George Chandler took pride in this move taken by the son who was ably following in his father's footsteps. Although by now he had been on the sidelines for nearly 10 years, George Chandler glowed with satisfaction as Belmont Hartland continued his winning ways under Chandler ownership until he had amassed a total of 19 grand championships in 1928 and 1929, including those awarded at Denver; Ogden, Utah; Portland, Oregon, and several Canadian exhibitions. Before Camden sold him, Belmont Hartland had been undefeated as a junior yearling, and winner of the junior championship at the 1926 American Royal, and the 1927 Denver and Ogden shows. He was, in fact, acclaimed by some observers to be the most perfectly formed beef bull of his day.

The matings of the descendants of Prince Domino 164th and Belmont Hartland, the first two major herd sires bought by Herbert Chandler, proved to be one of the happiest nicks in Hereford breeding of its era. From these crosses came a remarkable line of Chandler's Belmonts and Lady Hartlands. It was one of the latter which gave birth to the bull calf which was to mark the next Chandler step. She was Lady Hartland 54th which, mated with Donald Domino, cited above, in 1936 produced Donald Domino 16th.

One of the most successful showring winners of his time in the West, he proved to be a remarkably powerful sire, especially through his sons, Mark

*Herbert Chandler, left, and Charles Chandler comprised one of the few father-and-son pairs ever to serve as presidents of the American Hereford Association. Here they were pictured with one of the numerous prize-winners produced in the herd of Chandler Herefords.*

Donald and Donald Dhu, both from dams by Mark Domino by Prince Domino 164th. Thus again was demonstrated the efficacy of linebreeding to a prepotent herd sire. Mark Domino's dam, incidentally, was another product of the Chandler foundation program, she being a daughter of Prince Rupert 60th and her second dam having been sired by Beau Donald 33d.

It deserves to be noted, too, that Mark Donald's dam was the great Chandler show cow of 1937 and 1938, Miss Mark 34th, whose championships during a three-year campaign on the tanbark were climaxed by victories at Kansas City, Denver and Ogden. Her dam was a daughter of Belmont Hartland, and her second dam was a Debonair 26th daughter.

A further comment on the productivity of the herd's champion females is that Bonnie Catherine, Denver champion female of 1930, sired by a grandson of Prince Rupert 60th and out of a Debonair 26th cow, became the dam of one of the most noted of the Chandler's Belmonts. He was Chandler's Belmont 30th, which Bonnie Catherine dropped from a mating with Belmont Hartland.

Herds across the country are beneficiaries of the Hereford operations of successive generations of Chandlers, including Herbert's son, Charles, who took over with the passing of his father in 1973. Dozens of Hereford lines trace back to Chandler foundations, many of them still prominent at a point well into the final quarter of the 20th century. Among them might be cited the Sam Donalds, the Royal Saint line, the Big Arthurs, the descendants of TT McK Command 82d, most of the lines whose names involve the word Dhu, the Mark Arthurs, the Beau Marks, various Selkirk lines and others, on and on. In a survey based on the 1976 Herd Bull Edition of *The American Hereford Journal*, the Chandlers found that 224 advertisements represented herds which either featured or were using bulls and/or females with Chandler bloodlines in their inheritance.

Both George and Herbert Chandler were "personalities" in their own right, in addition to their activity as ranch and herd operators. For some 15 years after his active participation in the herd's operation ended, George Chandler continued to be one of the *Journal*'s ablest and most vigorous correspondents, giving a new generation of breeders the benefit of his constructive experience. Numerous articles conveying his observations and conclusions, and occasionally pithy comments which evoked responses from fellow breeders, appeared in the magazine well into the 1930's.

Herbert Chandler, on the other hand, is not remembered particularly for articles contributed to the *Journal* but, rather, for his verbal eloquence and good humor wherever Hereford enthusiasts gathered. So much so, in fact, that he sometimes was referred to as "the silver-tongued orator from Oregon," and on other occasions as "the bard of Baker." Both men were unforgettable characters in the strictest sense of the term.

An unusual distinction for George Chandler is the fact that both his son Herbert, and his grandson Charles, became presidents of the American Hereford Association. Herbert was elected to the post for the first time in 1933, and then was named again to the position in 1952, holding the post at the time the new Hereford Association headquarters building was dedicated by President Dwight D. Eisenhower in 1953. Charles was elected to the the post in 1961, being at the helm when the Association acquired *The American Hereford Journal* from Walker Publications, Inc., as its official publication.

And, of course, with new generations of Chandlers coming along, this family's Hereford story is not yet ended!

# Overton Harris

HARRIS, MISSOURI
1856-1931

In an era when visual appraisal, which in a later period came to be described as "eyeballing," along with pedigree, were the primary factors in developing a Hereford breeding herd, no Hereford establishment attained greater prominence or gained wider acceptance during the first quarter of the 20th century than that established by Overton Harris of Harris, Missouri.

As with several others who pioneered in the registered Hereford field, Overton Harris, whose entire lifetime was spent on the north Missouri farm where he was born in 1856, first was a cattle feeder, in association with his father. This was in the latter part of the 1880's and the 1890's. The father's distinct preference was for Shorthorns, then the most prominent and popular beef breed. Good steers possessing evidence of their good breeding were the rule in the Harris lots, but the senior Mr. Harris had no interest at all in becoming a producer of registered cattle of any breed.

The son was distinctly unlike him in this latter respect. It was his conviction that if well-bred grade cattle did well, a purebred registered herd would do better. And he had a breed preference which differed from his father's. Although he did not know a great deal about Herefords from personal experience, his ambitions tended toward the Whitefaces.

The father and son became partners in the farming and feeding operations when the son reached manhood, with the junior member's influence on operations expanding somewhat. However, he was unable to convert his father to the Hereford breed. But the opportunity finally came for the son to act. He took a positive step, with results which breed history records.

Along with the cattle-feeding operations at the Harris farm, a modest grade breeding herd also was maintained. The father's convictions in regard to quality led him always to use purebred bulls in this unit, and customarily to choose Shorthorns. But the son's insistence in behalf of the Whitefaces led the father to make a concession.

It happened that Gudgell & Simpson of Independence, Missouri, and James A. Funkhouser of Plattsburg, Missouri, had scheduled a combination sale on October 23, 1896, at Independence. The father agreed that Overton might attend that sale to purchase a bull for service in the grade herd. By earnest persuasion he also got the consent of his father to purchase one female at up to $150–and no more.

As was later recounted, Overton was filled to the brim with excitement as the sale opened. He knew good bulls, and soon found a Gudgell & Simpson entry in the ring that he especially liked, a son of Lamplighter, and bought him for $210. Then he set out to get a heifer at the $150 limit. Repeatedly he bid to his limit and had to stop. This went on for some time. The auctioneer, a leader in the Hereford field at that time, noticed what was going on and seemed to sense the situation. Finally, an especially nice heifer, a Gudgell & Simpson daughter of Druid by Don Carlos, entered the ring and as usual Overton bid his $150. Instantly, the auctioneer said "Sold!" Overton later said that he almost went into shock, but he quickly recovered his equilibrium and in his enthusiasm, without paternal authority, proceeded to bid off three more heifers, two of them at $165 each and the other at $140. Two each of the four came from each selling firm's consignment.

After some worry about the problem of making settlement for his purchases, a happy thought came to Overton's mind. He would give two checks for the cattle–one partnership check for the bull and one heifer, and his personal check for the other three heifers. The father said nothing about the extra purchases until the end of the year when accounts were being settled up. Noting then that Overton had given his own personal check for the extra heifers, he informed the son that he could not engage in a side business, and if they were to be partners they would have to be partners all the way.

But still the father thought they were outrageously high-priced, and that their cost above beef prices would be a dead loss. He didn't change his mind until two or three years later when two men from Texas turned up at the farm and gave Overton a check for $2,000 for five heifers. Those five heifers had

*The famous round barn marked the headquarters for the Model Herefords of O. Harris & Sons. It long was a landmark to which Hereford breeders from across America were drawn to inspect the herd and to buy of its output privately and at auction–Risk photo.*

made more money than all of the steers fed on the farm that year, and from that day on the senior Mr. Harris was completely converted to registered Herefords. He told Overton to go the limit and he would back him.

From that point onward, Overton Harris, or "Obe" Harris as he came to be generally known to his friends, made registered Herefords his Number One priority. He soon purchased the bull, Benjamin Wilton, and 20 young cows, all but one with calf at foot, from an adjoining neighbor, John E. Stone. All represented the breeding of Cornish & Patten, at Osborn, Missouri, and it was David P. Cornish of that firm who advised him to make the purchase. Already possessing a love for Herefords, this acquisition provided the second of the essentials for success: the right kind of foundation material. He was now on his way–with success to a high degree coming even faster than an optimistic and enthusiastic Overton Harris could possibly have anticipated.

One of the cows in the Stone purchase, Betty, a granddaughter of Anxiety 4th, at the time Harris bought her was carrying a heifer calf by Benjamin Wilton. This calf, named Betty 2d, compiled a show record which included grand championships at numerous fairs and expositions, including the American Royal and International shows in 1901. Thus, the Harris name almost immediately became one to reckon with. Betty 2d then went through the Chicago International auction ring at a record price of $4,500 to J. C. Adams of Moweaqua, Illinois. Betty 2d's success, along with that of her mates, fueled the Harris inclination toward the showring with the result that the Harris herd in remarkably short order became one of the breed's most successful. Most of the Harris entries in these earlier years were by Benjamin

Wilton, the bull purchased from Stone, which was a descendant of Garfield on his sire's side and out of a dam by Anxiety 4th, these two bulls being the premier breed builders of their time.

An extraordinary Hereford show, hailed as the greatest ever staged up to that time, came in 1904 at the Louisiana Purchase Exposition in St. Louis, Missouri. Harris entries competed there against representatives of most of the leading herds of the period, and proved to be sensational winners, including five firsts and five seconds, plus the female junior championship. Two gold medals also were won–one for the exhibitor who won more prizes than any other exhibitor on cattle of his own breeding and the other for the exhibitor whose cash winnings exceeded those of any other exhibitor.

A recital of Harris winnings from then on for at least the 15 following years would require pages. An idea of their extent is provided by a statement in a 1914 Harris auction catalogue which stated: "At 14 of the leading shows of 1911 and 1912, our herd won 122 first prizes out of a possible 210, and 54 championships out of a possible 84–more than were won by all other competing breeders combined." At the American Royal alone, Harris bulls accounted for 20 of the 27 championships awarded in the nine shows they entered between 1908 and 1918.

Contributors to the herd's ancestry by this time were other sires, including Beau Donald 5th, which had been bought as a successor to Benjamin Wilton. Bred by W. H. Curtice of Eminence, Kentucky, this bull was sired by old Beau Donald while his dam was double-bred to Earl of Shadeland 9th. A state fair champion himself, Beau Donald 5th sired numerous winners for Overton Harris, but even more significantly he produced scores of cows which contributed to the prestige of the Harris breeding herd. As late as 1920, it was said, too, that every herd bull then in service was strongly bred to Beau Donald, usually through Beau Donald 5th.

For example, that greatest of all Harris herd bulls, Repeater, was sired by a grandson of Beau Donald 3d by Beau Donald, and thus a half-brother to "the 5th." At the Chicago International of 1908, Harris saw a yearling bull in the string of E. W. and A. M. Heath of Smithboro, Illinois, that he considered a sensational prospect. This bull was Repeater. Of incidental interest is the fact that his dam represented the Hesiod family which rose to prominence in the herd of James A. Funkhouser, the man from whom Overton Harris bought two of those first four heifers.

Repeater was bought primarily as a show bull, having headed his strong senior calf class at that 1908 Chicago show. The new owner was reluctant at first to use him as a sire because of a small feather of white on his loin, and the veteran showman John Letham even had thought of buying him to alter and show as a steer. This is a token of his apparent sterling individuality as a calf since the Scottish-born Letham's perspicacity in anticipating bovine development was legendary. W. S. Van Natta, the noted pioneer breeder of Fowler, Indiana, also had considered the calf as a future Van Natta herd bull. Later, after the Harris purchase mainly for showing, Van Natta persuaded the Missourian to try Repeater on 10 head of cows. The resulting calves included Repeater 7th and Miss Repeater 11th, two of the most famous of all Harris winners, and, incidentally, the two whose picture adorned the front cover of Alvin H. Sanders' early breed history, *The Story of the Herefords*, published in 1914.

Not only did Repeater win grand championships for Harris at Kansas City and Chicago, plus at least a dozen leading state fairs, but his sons and daughters proceeded to do likewise in their turn, to the extent that the Harris herd soon was unsurpassed in the ranks of America's leading show herds. Repeater Jr., a son of Repeater, was grand champion bull at the American Royal three consecutive years, the only Hereford bull ever to accomplish this feat. This bull and two other sons of Repeater won for Harris at Kansas City's American Royal the Pereda Cup, a sterling silver trophy offered by Celedonio Pereda of Argentina for the best three bulls bred and owned by exhibitor, retiring the Cup from competition in 1920 after gaining permanent possession by winning it three years in succession.

Likewise, the Harris firm, by then known as O. Harris & Sons, and comprising in addition to the founder his son, A. W. (Wood) Harris, and two sons-in-law, Gird McCullough and O. H. Moberly, retired the Duggan Cup, a silver trophy offered by the Duggan Brothers of Argentina for the best two animals, bull and female, bred by the exhibitor, by winning it for the second time at the 1920 Chicago International.

Not only did the Model Farms Herefords of the Harris firm win sensationally but buying demand for them was logically high. Annual auction sales were highlight events, with averages well up in four figures being recorded, and with individual auction prices ranging up to $35,000.

Gay Lad 6th, a member of W. S. Van Natta's Prime Lad family, joined the Harris sire battery not long after the advent of Repeater, and was the only bull to defeat the latter in championship competition during the 1910 campaign. Just as they were show-ring contemporaries, they likewise were the founders of the families–the Repeaters and the Gay Lads–that put the Harris name in the breed history books to stay.

It is perhaps relevant to cite that one of Canada's earliest Hereford families to gain international fame was founded by two sons of Gay Lad 6th. They were Gay Lad 16th and Gay Lad 40th, both successful show bulls. Under the ownership of Frank Collicutt of Willow Spring Ranch at Crossfield, Alberta, they founded the WS Gay Lads and WS Gay Lasses which for years were the most widely popular Hereford family in western Canada, if not in the entire Dominion.

Overton Harris practiced what he preached, and the results offer proof of the rightness of his philosophy. He once wrote the following, as indicative of his beliefs:

"To a beginner my advice is to buy good cattle. There is always room at the top, and, always room to make improvements in the herd. If you buy cheap cattle you will always be five years behind. Buy good cattle and keep going. If you buy the cheaper class, one that will do 'just as well,' you will spend five years getting them to equal those which are the good kind now–and where will those which are the good kind now be five years hence. You can't keep up unless you start right."

The farm and livestock depression of the middle 1920's hit the Harris operations, as it likewise did a great many others. Then, in 1931, the herd's founder died. His son, A. W. Harris, and the latter's son, Anderson W. (Andy) Harris, carried on until the final O. Harris & Sons dispersion sale was held in 1937.

Both Overton Harris and A. W. Harris served as presidents and members of the board of directors of the American Hereford Association. The former was at the helm in 1916, and the latter headed the organization in 1935. Overton Harris again was serving on the board at the time of his death.

*No herd was more successful at the major shows in its heyday that that of O. Harris & Sons of Harris, Missouri. This trio comprised the "best three bulls" in a special competition for the Pereda Cup (shown), awarded at the 1918 American Royal by a famous Hereford-breeding family in Argentina. The bulls, left to right, are: Repeater's Model, Repeater Jr., and Repeater 129th. Repeater Jr. was the only bull ever to become grand champion bull three successive years at the Royal, which he accomplished in 1918, 1919 and 1920. Nor, in view of later age restrictions for show cattle, is this feat ever to be duplicated. The Harris firm won this trophy for three successive years at the Royal, and thus retired it–Risk photo.*

# Murdo Mackenzie

DENVER, COLORADO
1850-1939

Some cattlemen contributed notably to Hereford history not through their success in breeding registered Herefords but in demonstrating to an outstanding degree the breed's worth in commercial beef production. Such a man was Murdo Mackenzie, a native Scot from far-north County Ross, who came to America in 1885 at the age of 35. During the ensuing 54 years he left an indelible mark upon the range beef-cattle industry through his service as manager of two of the most famous of all range outfits.

The first of these was the Prairie Cattle Company, which operated mainly in the Texas Panhandle, and, after around six years, at the helm of the vast Matador Land and Cattle Company. The latter Scottish-owned concern operated on a huge scale in both the upper and lower Texas Panhandle, with later extensions under Mackenzie's guidance reaching northward as far as South Dakota and Montana, and onward, at last, as cow-country historian J. Evetts Haley once stated, to "an ocean of grass on the plains of Saskatchewan." In this era of expansion, the same writer hailed Mackenzie as "one of the sturdy symbols of the West."

When Mackenzie took charge of the Matador the herd consisted of some 65,000 cattle, all of the ordinary sort and of varied colors, since grade bulls descended from all three of the principal beef breeds had been in service. But

Mackenzie had no use for mongrel cattle, and as he continued his study of the improved breeds which had been begun during the years when he managed the Prairie outfit, he reached a firm decision to breed from purebred bulls only, and, at the same time, to cull out as quickly as reasonably possible all inferior cows, beginning immediately with those at the lowest end of the scale.

The first registered Hereford bulls bought for the Matador came in 1892 from the good herd of Fowler & Tod of Maple Hill, Kansas, in which the Gudgell & Simpson influence was paramount. At the same time, Mackenzie also purchased some purebred Shorthorn bulls, but after an adequate trial a firm decision in favor of Hereford bulls exclusively was made. The Herefords were highly successful in the test, the improvement in conformation and quality of the resulting calves being so apparent that the choice was never in doubt.

But Mackenzie now was faced with a dilemma. He was not the owner of the Matador, but only the manager of a company owned by its Scottish stockholders. He knew that to supply a herd of many thousands of cows with a sufficient number of good purebred Hereford bulls would be a very expensive process. He thus adopted a policy of buying purebred cows and holding them on a separate range for the purpose of breeding on the Matador at least a part of the bulls required. This was an undertaking in which Mackenzie was successful. Old-time rangemen had thought such a program in a semi-arid and sparsely grassed range region would be a failure.

After an adequate trial, Mackenzie reported:

"If the Hereford cow is supplied with a sufficient amount of grass she will produce a calf as regularly as the cows on Missouri and Iowa pastures, and at four years old the bulls they produce will have grown out to be just as large and of as good conformation and quality as any you will find in the grain-growing states. All that is required is care and the culling out of the cows undesirable for breeding; but not only is this care required on the range, but it is required on the farm as well if one expects to get the best results.

"I do not wish the breeders of other purebred cattle to feel that I have any prejudice against the other breeds; all breeds to my mind are good in their own place, but in large pastures where cattle have to hustle for themselves I have no hesitation in stating that my experience has been that the Hereford has it over them all."

Shortly the Matador purebred Hereford herd was producing around 150 bulls each year to go into the commercial units, and prices up to $1,000 were being paid for good registered bulls which sired those improved range bulls. This number was supplemented by purchases from outside sources at prices of up to $250 a head of bulls for commercial herd service, which was well

*Murdo Mackenzie on a walkway at the Chicago stockyards where a load of Matadors had been adjudged champion yearling feeders at the 1902 International Live Stock Exposition.*—Breeder's Gazette photo.

above the average range-bull figure for that time. His aim was perfectly clear.

Murdo Mackenzie stated in a 1918 interview that in the grade herd none but registered Hereford bulls had been used for some 20 years. So, for all practical purposes, he went on, "it is a purebred herd." Sires were in use in the grade herd that are good enough to head most purebred herds, he added.

It was Mackenzie's belief that no bull was too good for range use, and he strongly resented the then prevalent idea of some that the range was an ideal "dumping ground" for bulls considered to be not good enough for registered herd service. His firm conviction was that only bulls of the best bloodlines, big in bone with plenty of scale and fine fleshing qualities, should be allowed to perpetuate the Hereford breed on the range. When Mackenzie found bulls which met his requirements, he backed up his words with a willingness to pay the price required to get them regardless of the source of the competition.

So it was that Mackenzie became one of the breed's strongest proponents, and his stature in range industry councils obviously influenced other cattle producers as they observed the Matador results and heard him speak out for the Whitefaces, as, for example in this 1914 comment: "In this enlightened age, everybody knows that Herefords cannot be equalled as range animals, and we have found that the nearer purebred they are the better they do."

A factor which heightened Mackenzie's enthusiasm for Herefords, almost from his beginnings with them, was the breed's prolificacy. The goal of high-percentage calf crops held great appeal for him, as a range herd manager, and shortly after he assumed the helm and turned to Hereford bulls exclusively four out of five Matador cows were producing calves annually. These 80-percent crops were considered remarkable under the harsh, open-range conditions that prevailed—and the percentage rose to even higher levels as time passed.

At one time, Mackenzie was prevailed upon to give a trial to a group of bulls of another breed—not specified in his recollections. But he did recall that the calf crop in the pasture served by those bulls dropped by 40 percent. A prompt return to Hereford bulls was quickly reflected in a return to the normal higher-percentage.

So it was that Murdo Mackenzie, generally considered by range-industry students as the most forceful and successful large-scale operator of his era, from headquarters offices first at Trinidad, Colorado, of which he once was

*Always searching for the means to further improve the feeder cattle output of the Matador Land & Cattle Company, Murdo Mackenzie purchased this grand champion load of yearling bulls from the noted herd of A. B. Cook of Townsend, Montana, at Denver's National Western in 1921. The price paid for these and 17 others of like breeding and age was $500 a head in a year when the average price paid for all registered Hereford bulls was only $215.*—Rocky Mountain *photo.*

mayor, and later at Denver, ruled operations on the Matador for more than a generation. That span continued until his death in 1939, after which his son, John, who had long assisted his father, assumed the management post. The latter continued at the helm until the Texas holdings, then totaling 812,000 acres, and 47,000 high-class commercial Herefords, were sold in 1951. This sale to American purchasers marked the end of the era when vast western ranches were under foreign ownership. It was the last of its kind.

The conformation and quality which the Matador's registered unit attained, even in its earlier years, is indicated by several significant facts. For example, at the 1911 National Western Stock Show in Denver, the Matador company exhibited the senior champion bull and the junior and grand champion bull, although showring competition seldom if ever again was undertaken. This initial effort was tried in order to see how the Matador bulls stacked up against the entries of prominent registered breeders.

Many bulls bred by Gudgell & Simpson were used. One of them, Publican, the sire of Domino and grandsire of Prince Domino, was brought back from the range after a period of service on the Matador by Robert H. Hazlett of El Dorado, Kansas, for whom he founded an important Hereford family. Obviously he left many descendants on the Matador range.

Matador paid the $3,000 top price in the 1928 International sale at Chicago for a bull consigned by R. P. Lamont, Jr., of Larkspur, Colorado, and this sire's caliber later was indicated when a son became junior and grand champion bull for Lamont at the 1929 International Live Stock Exposition, and later headed the herd-bull class at the Golden Anniversary Hereford Show at Kansas City in 1932.

In 1931, and again in 1935, Matador bought the top-price bull of the National Western sale in Denver, the latter also having been the champion bull of that year's Denver show under the banner of Foster Farms of Rexford, Kansas. Several bulls of Anxiety 4th ancestry were bought from P. J. Sullivan of Wray, Colorado, whose herd was one of the leaders in its heyday. For example, in Sullivan's 1927 fall sale, the Matador bought 10 of the 12 older bulls in the auction, including the event's top seller at $1,800, a son of Superior Mischief for which Sullivan paid Mousel Bros. $22,000 in 1920, the highest auction price ever recorded by the Mousel firm. On another occasion Murdo Mackenzie bought from Chandler Herefords of Baker, Oregon, a high-ranking carload of bulls in the Denver show, representative of some of the breed's best bloodlines. An earlier carlot purchase at Denver came in 1921 when he bought the show's grand champion load of yearling bulls, plus 17 others—largely of Beau Brummel ancestry—from A. B. Cook of Townsend, Montana.

This sketchy survey is by no means complete, and is intended merely to

*As finished beeves the Matadors always displayed their strong Hereford ancestry. This load, bred on the Matador's Texas range, was fed by Dan D. Casement at Manhattan, Kansas, who exhibited them to the carlot show's grand championship at the 1931 American Royal.—Morton photo.*

indicate Mackenzie's activity in infusing the influence of some of the most prepotent of the breed's top families into the Matador herd. And, although he always sought quality and what he deemed to be desirable conformation, he was, indeed, also bloodline-conscious. In an era when some onlookers theorized that buyers of registered Herefords were paying too much attention to pedigrees, Murdo Mackenzie once was publicly complimented in a livestock publication for bidding on bulls in a National Western sale without even glancing at the pedigrees in the sale catalogue. Mackenzie responded immediately that he *never* bought a bull without being fully aware of his pedigree, and pointed out that on the occasion to which reference had been made he had previously studied the ancestry of every bull in which he was interested. He believed in both individuality *and* pedigree, he said, which may have been the main reason the Matadors won such wide acclaim for such a long period under his management.

This acclaim frequently was manifested in the winnings of the Matador-bred steers in carlot competition at major stock shows. The ranch itself successfully exhibited carloads of feeder cattle in such contests as that at Denver where its winnings ranged up to the grand championship. Matador steers finished in the feedlots of such widely known cattle feeders and exhibitors as Dan D. Casement of Manhattan, Kansas; John Imboden of Decatur, Illinois, and others acquitted themselves so creditably in winning purple ribbons as finished beeves in major stock-show competition that none could doubt the influence of their top Hereford inheritance for generation after generation. Probably no listing has ever been made of the carlot show

*Matador steers on feed in the lots of the Harris Grain and Cattle Company at Sterling, Colorado, always were of market-topping caliber as finished beeves—a tribute not only to good feeding but even more importantly to their beef-producing ancestry.*—Record Stockman *photo.*

winnings of the product of various commercial cattle herds, but if there had been Matador surely would have ranked among the very tops.

These performances merely were demonstrations of what the Matadors could accomplish. Dozens of commercial cattle feeders bought the Matador calves and yearlings year after year to finish out for the beef market, with results that usually were gratifying all around. This was what Murdo Mackenzie had in mind almost from the beginning of his long tenure at the head of Matador operations.

After some 20 years with the Matador, Mackenzie left for an 11-year period during which he was general manager of the Brazil Land, Cattle and Packing Co., Sao Paulo, Brazil, a syndicate which held 10,000,000 acres of ranch lands and operated a herd of some 300,000 cattle, plus other kinds of livestock, in South America. His Hereford advocacy continued in his new post, as indicated by a *Hereford Journal* news story in 1914 announcing his selection and purchase in the United States of some 450 Hereford bulls to go to the Brazil outfit. W. J. Tod of Kansas, prominent Hereford breeder and commercial cattleman, said that these bulls were "just about the top of any bunch the same size" that he had seen in a long time. The Brazilian venture collapsed after World War I, and by 1922 Mackenzie was back on the job at the Matador's helm.

And "back on the job" meant exactly that. A contemporary account in a newspaper as he reached his 85th milestone cited his sturdy nature and unremitting devotion to business duties in these words: "Despite the heaviest snowstorm of the season, he was at his desk in his Denver office at 8 A. M."

This was not all, as the account continued: "In the evening Mr. Mackenzie entertained a large number of friends at a dinner party in his home, where he made a grand attempt to extinguish at a blow the 85 candles placed upon the birthday cake. Then his beloved violin, and a rendition of the Highland Fling—for that 'fiddle' has long been an integral part of its owner."

Mackenzie's prominence in ranching circles, and his strong influence in that field, made his advocacy of Whitefaces of maximum value to the Hereford breed. He had been president of the Texas and Southwestern Cattle Raisers Association as far back as 1901, continuing in the post for three terms. He served ably in 1906-7, and again in 1911, at the helm of the American National Live Stock Association, which years later became the National Cattlemen's Association. He was elected to the board of directors of the American Hereford Cattle Breeders' Association in 1910.

His rugged individuality, keen mind, tenacity and knowledge of cattle made him a leader of outstanding prestige and influence, and it was to the Hereford breed's great advantage that its performance under his guidance on the Matador range led him early to become one of its strongest advocates. Where he led, hundreds of others followed. By word and example, he contributed vastly to Hereford advancement throughout the ranching country.

All who had known him, either in person or by reputation, were gladdened when, in 1981, he was honored by admission to the Hall of Fame of Great Westerners at the Cowboy Hall of Fame and Western Heritage Center at Oklahoma City. It was a signal recognition which has come to only a few, and in honoring Murdo Mackenzie this prestigious institution, in essence, also honored itself and the Hereford breed.

ROBERT D. MOUSEL

HENRY L. MOUSEL

# ROBERT D. MOUSEL

CAMBRIDGE, NEBRASKA

1877-1967

# HENRY L. MOUSEL

CAMBRIDGE, NEBRASKA

1878-1960

COMPRISING the landmark Hereford-breeding firm of Mousel Bros. of Cambridge, Nebraska, for more than 60 years were Robert D. Mousel and his year-younger brother, Henry L. Mousel. Building on foundation stock from the herd of Gudgell & Simpson of Independence, Missouri, they gained national renown as perpetuators of the Anxiety 4th bloodlines and of the linebreeding practices used by the historic Missouri firm. Mousel Bros.' herd, indeed, became famous as the fountainhead of straight linebred Anxiety 4th Herefords after Gudgell & Simpson's dispersion sale in 1916, and it was said when Robert and Henry Mousel died during the 1960's that they probably were the last remaining Hereford breeders to have dealt directly with the owners of that noted herd.

Few Hereford breeders who gained fame as a result of their Hereford endeavors began as inauspiciously as these Nebraska brothers who were born in a sod house constructed on the 160-acre homestead of their father, Michael

Mousel, a native of Luxembourg, who had emigrated to the United States at the age of 19. Their initial acquaintance with Herefords came with their father's use of the first Whiteface bulls ever seen in that region in his grade herd of mostly Shorthorn blood. The performance of those bulls, and the substance and quality of their calves, made a lasting impression on the boys as they helped their father with the chores. The direction of their lives was set.

In the late 1890's, when Robert was 21 years old, and Henry a year younger, the firm of Mousel Bros. came into being when the two young men responded to an advertisement they had seen in *The Breeder's Gazette*. This resulted in their mail-order purchase of two purebred Hereford heifers from D. P. Williams & Son of Guthrie Center, Iowa, for $100 each–a sum that almost completely depleted the treasury of the new firm. The boys bred their heifers to registered Hereford bulls which their father had bought in 1896. A year later they bought another Williams heifer, and that same year 11 heifers were purchased from C. B. Andrews of Fort Collins, Colorado. Mousel Bros., as a firm, was definitely on the way, with their foundation stock tracing mainly to Garfield and Lord Wilton.

A turning point came for them in 1903 when their ambition led them to attend the American Royal show in Kansas City. For the first time they saw there samples of the Gudgell & Simpson output, and what they observed had a lasting effect on the subsequent operations of Mousel Bros. Considerable time was spent around the Gudgell & Simpson stalls and at others where the bloodlines of that herd were featured. But as beginners, with only modest financial resources, they found it necessary to temper their enthusiasm with good judgment. The result was that they bought for $255 from Steele Bros. of Richland, Kansas, the bull Princeps 4th. He was not entirely of straight G. & S. breeding but both his sire and dam were by grandsons of Anxiety 4th.

Although Princeps 4th stayed in their herd only a short time, among his get he left two bulls which ranked high in their class at the 1907 American Royal and the 1907 and 1908 International shows. In fact, at the 1907 International against representatives of many leading herds these two stood first and second in their class. Many were surprised to learn in the fall of 1905 that the Mousels had sold Princeps 4th to the well-known firm of Cargill & McMillan of LaCrosse, Wisconsin, for the relatively long price of $1,750. The reasons, as related many years later by Henry Mousel, was that the brothers "needed the money." But the deal also brought a measure of recognition to the Nebraska firm when Princeps 4th became senior champion at the 1906 International for his new owners, and when one of his daughters won the similar honor in the female division of the Chicago event a little later.

Mousel Bros., even before letting Princeps 4th go, hadn't lost any time getting into the selling end of the Hereford business. They had their first public auction in 1904, in which their bull offering averaged $117.70, a fair figure for the times. Four years later they held a dispersion sale for the purpose of letting one person who held a small interest in some of the cattle get out. The distance the Mousels already had come in establishing themselves as breeders of high-class Herefords is evident in the fact that one of their show bulls, Alto Hesiod, brought the top price of $1,050, and the son of Princeps 4th that had stood second in class to his half-brother at the 1907 International was sold for $900.

Another reason for the dispersion sale was the fact that by now the Mousels had firmly decided to concentrate their efforts on the production of linebred Anxiety 4th Herefords and to follow as nearly as possible in the footsteps of Gudgell & Simpson. During their earlier years with Herefords, they had never ceased admiring the G. & S. cattle, and now, they decided to act.

One of their first moves was to acquire the Gudgell & Simpson-bred Beau Mischief. This son of Beau President by Beau Brummel and out of the cow Mischievous by Lamplighter, had been bought as a calf in 1907 for $500 by J. A. Larson of Everest, Kansas. Larson carried him along for showing at that fall's American Royal but the youngster was not in top show condition and ranked fourth in class, with a Mousel entry ranking a notch above him. But nevertheless the Nebraskans were greatly impressed by the Larson bull, Beau Mischief, and begged the Kansan to put a price on him. Larson refused.

But via the grapevine "wireless" they later heard that Beau Mischief had not done well at the Larson farm and was being turned back to Gudgell & Simpson as a "shy" breeder. He was to be shipped to the G. & S. ranch at Edmond, Kansas, located not far south of Cambridge. Never

*Henry L. Mousel, left, and Robert D. Mousel found a moment for relaxation on a bale of straw before the start of the Mousel & Coder sale at Cambridge on October 16, 1954. Don Coder, who raised Herefords at Wellfleet, Nebraska, was a son-in-law of Henry Mousel.—*Hereford Journal *photo.*

having lost their desire to own the bull, the brothers decided to go to Edmond to see him. The result was that Mousel Bros. became his owners for only $150. But it was a gamble.

As Henry Mousel later was to say, "He was a sorry sight. He was sick, sore, lame and disordered, but he still looked like a herd bull to Bob and me." Preferring not to hear the adverse comments of their neighbors about their new bull that had been so badly foundered that he walked only with difficulty, they slipped him in to Cambridge in a boxcar, loaded him into a high-sided wagon and hauled him to their ranch. Believing that he had not been let down properly after showing, they turned him out onto the springy turf of a buffalo-grass pasture, and awaited his recovery.

Beau Mischief came back slowly. He sired only a few calves the first year, but those few convinced the Mousels of the bull's prepotency and made them more than ever determined to succeed in restoring him to full service. The following year he began to breed consistently and was active until his death in 1917, siring 222 sons and daughters for Mousel Bros. Ten calves by him had previously been recorded by Larson, and three bulls and two heifers by Gudgell & Simpson, the latter five in the interim between his return by Larson and his purchase by the Nebraskans. An interesting sidelight is that one of the three bulls he sired for Gudgell & Simpson was Beau Blanchard, later to bring fame to Jesse Engle & Sons of Sheridan, Missouri.

Beau Mischief's contribution to the Mousels' success is incalculable. Sons and daughters which were sold in their auctions, beginning in 1912, brought a total of almost a quarter of a million dollars. Others were sold at private treaty. And, of course, scores of them were retained in the Mousel breeding herd. His highest-priced son was Superior Mischief which brought $22,000 from P. J. Sullivan, then of Wray, Colorado, in the record-breaking Mousel Bros. sale of 1920. That auction established a new high mark for the breed when 73 head averaged $4,020.

If their acquisition of Beau Mischief was the first master stroke of the Mousels, their second came a few years later with their purchase of 92 head of their selection from the Gudgell & Simpson herd only a few months before that great herd's final dispersal in 1916. Charles Gudgell by then was the only survivor of the firm's original members, and he was of advanced years.

With an awareness of Mousel Bros.' determination to pursue a breeding program similar to that which had brought fame to the Missouri firm, permission was granted to the Mousels to select this group comprising many of the herd's best. Final selections included 30 cows with calves at side and 32 heifers. The cows included Donna Anna 22d, the dam of Domino. The entire group contributed predominantly to the foundation for what came to be

*This is probably the last picture ever made of the history-making Prince Domino Mischief. It was taken in early December, 1924, when he had just passed four years old. He posed here with his first son, Advance Mischief, destined also to become a history-maker, on a bridge over Medicine Creek. Herdsman Bill Hendry, Jr., was at the halter. Prince Domino Mischief died as the result of an infection on Christmas day, 1924.*

generally recognized as Mousel Bros.' fountainhead herd of straight-bred Anxiety 4th Herefords, and "Bob" and "Hank" Mousel were widely heralded as the logical successors of Gudgell & Simpson.

It was a truly great daughter of Beau Mischief, out of a dam which was included in this final purchase from Gudgell & Simpson, that did more than any other of Beau Mischief's get to lastingly establish the Mousel name in breed annals. This heifer was Mischief Maker 6th, second-prize winner in her class at two consecutive Chicago Internationals, and her dam, Gwendoline 34th, was hailed by Charles Gudgell as the greatest of the daughters of Beau President.

Shortly after her last International appearance, arrangements were made to send her to the ranch of Fulscher & Kepler of Holyoke, Colorado, for mating with Prince Domino, son of Domino, which by then was making a name for himself as one of the all-time greatest of Gudgell & Simpson-bred herd bulls. The Colorado firm had bought him five years earlier and his remarkable prepotency had enabled him to sire such a notable array of sons and daughters that both the Mousels and Otto Fulscher were convinced that such a mating was certain to produce an outstanding result.

It did. The calf gained fame as Prince Domino Mischief. He was deemed by his owners to be of herd-bull caliber from the outset, and a year or two later in a sale catalogue footnote they said: "It will be observed here that we have reduced to one individual the blood of the best individuals in this great family of Herefords."

A few weeks after that catalogue footnote appeared, the Mousels sent Prince Domino Mischief to the 1923 National Western show at Denver where, despite a great age handicap, he won first in the senior yearling bull class. He was then retired to herd service. Although he died prematurely as the result of an injury after siring but two short calf crops, his descendants became major contributors to Hereford progress throughout the 20th century. Seldom, if ever, has a bull built so great a reputation on such a small number of calves.

Among his descendants are such families as the Advance Mischiefs, Superiors, Husker Mischiefs, Lamplighters (including the polled branches of the family), Prominos and Promises, Advance Dominos (again including such polled lines as the Gold Mines, Banner Dominos, Mesa Dominos and others), Champion Dominos and others. The dam of the family-founding Prince Domino Return was double-bred to Prince Domino Mischief. The founder of another family, the Zato Heirs, likewise was out of a Prince Domino Mischief dam, through Advance Domino.

From the viewpoint of the latter years of the 20th century, special attention is directed to Advance Domino 13th, a son of Advance Domino whose dam was a granddaughter of both Beau Mischief and Prince Domino. Fred C. DeBerard of Kremmling, Colorado, bought him as a calf for $1,000. In the DeBerard herd, this grandson of Prince Domino Mischief sired, among numerous others, Advance Domino 20th and Advance Domino 54th, which at the U.S. Range Livestock Experiment Station at Miles City, Montana, became the foundation sires of the Line 1 Herefords which achieved nation-wide prominence and popularity by the 1970s.

It is lastingly significant that no fewer than 30 pages of the 107-page *Hereford Lineage*, a 1978 genealogical listing of Hereford sires in primary service in American Hereford herds to that date, were devoted to descendants of Beau Mischief and Prince Domino Mischief, reflecting the Hereford-breeding genius of Robert and Henry Mousel. Their place in breed history is secure.

Additionally, Cambridge remained as of 1981 to be a name on the Hereford map, especially through the continuing Hereford-breeding activities of R. Wendell Mousel, son of Robert, and George Mousel, son of Henry. And who knows what Mousels may come to the fore in years to come?

# JESSE ENGLE

SHERIDAN, MISSOURI

1850-1926

As with numerous other 19th century beef producers, Herefords "sold" themselves to Jesse Engle of Sheridan, Missouri. A native of Kentucky who was only eight months old when in 1851 he and the other members of his family arrived in northwestern Missouri, he grew up in an environment of pioneer hardships. His father, who had only $65 in money when he reached what then was a frontier region, promptly pre-empted 160 acres of land, for which he eventually paid $1.25 an acre, and grubbed out 10 acres of it. The area as yet was so sparsely populated that for a considerable period the only market the Engles had for the corn produced on those 10 acres was afforded by Indians passing through the country.

When the son reached the age to begin doing for himself, he personally rented some land, engaged in farming, as had all of his known ancestors before him, and fed a few cattle. The only improved cattle which by then had penetrated into that rather remote area were Shorthorns, and young Jesse Engle was impressed by how well they performed their beef-making function in comparison with the native mongrels.

He prospered to the extent that by 1880 he was able to buy 110 acres in what is now Worth County, Missouri, where he, his wife and the earliest of their five children concentrated more and more on cattle raising. He raised some of the steers he fed, and bought others, and for some 20 years continued to favor Shorthorns, using grade Shorthorn cows which, mated with purebred Shorthorn bulls, yielded a satisfactory market product.

Jesse Engle's activity in this field gave him every opportunity to test and compare in his own lots not only the output of his own Shorthorn herd but also the steers which he purchased for feeding purposes from the viewpoint of their essential qualifications in producing beef economically. It was through the inclusion of a few grade Herefords in some of the bunches of steers he bought to put into his feedlot that his attention was first attracted to the beef-making capacity of the Whitefaces.

He had heard and read bits and pieces about this newer breed which was penetrating the country westward, but it was the actual demonstration of their potential in his own operations, and in competition with his own pure Shorthorns, that led him to experiment further. This he did through the purchase, around 1894, of his first Hereford bull from a neighboring farmer.

*"Uncle" Jesse Engle, as he came to be widely known in his latter years, was pictured here at the halter of one of his favorite show heifers, Belle Blanchard 112th, which was junior champion female at both the American Royal and International shows in 1922.—Smith & Morton photo.*

This bull was mated with a bunch of the Shorthorn cows, with results that were pleasing in the extreme.

Years later Jesse Engle commented that the experiment yielded steers of a better type, greater uniformity and possessing easier feeding qualities than ever before had been produced in his herd. The calves from this cross, he said, attained a marketable ripeness in from 60 to 90 days less time than any lot of steers he had previously fed, and topped the market when sold.

Market buyers of Engle steers in following months continued to demonstrate their liking for those sired by Hereford bulls. For instance, a buyer picking 15 tops out of 40 head of three-year-olds of mixed Shorthorns, Herefords and Angus chose 14 Herefords. On another occasion, Jesse Engle fed 90 western calves of mixed breeds until they were 18 months old. Forty head were then topped out for the Kansas City market. Those which were ready to go to market first were practically all Herefords, and weighed an average of 1,120 pounds.

In December, 1901, the Engle die was firmly cast for Herefords. He was convinced beyond a shadow of a doubt by his personal experiments and his keen eye as an observer. So, as the old year ended he made a new beginning with the purchase of his first registered Hereford females—three yearling heifers from the herd of Cornish & Patten of Osborn, Missouri. This trio

*The Gudgell & Simpson-bred Beau Blanchard, son of Beau Mischief, established a prominent Hereford family of his own through his years of service in the herd of Jesse Engle & Sons.—Risk photo.*

comprised the foundation of the Engle Hereford herd which rose to national prominence during the next dozen years or so. Two of them were by the sellers' noted imported herd bull, Weston Stamp, while the other one was by their Gudgell & Simpson-bred Boatman by Don Carlos.

It was Engle's good fortune in 1903 to acquire the imported bull, Ross, which had been bred in the R. T. Griffiths herd, one of England's leaders. Representing bloodlines then prominent in the breed's homeland, Ross had been purchased for import by K. B. Armour of Kansas City, who sold him to A. R. Garrett of Hopkins, Missouri, from whom Engle bought him.

As related by a contemporary of Jesse Engle, "Ross was a magnificent specimen. He had the size, the smoothness, wonderful bone and scale that Mr. Engle wanted. . . . His ability as a sire of smooth, even-fleshed calves of uniform type soon attracted attention, and Mr. Engle had no difficulty in disposing of his bull calves. And it was only through the prohibitive pricing of the heifer calves by Ross that he was able to retain the pick of them as future Engle brood matrons."

In 1911 came Beau Blanchard, the bull that was destined to make secure for all time the Engle name in breed history. At the Denver show of that year, Frank O. Gudgell, son of Charles Gudgell of the firm of Gudgell & Simpson, told Milton and Byron Engle, who along with their younger brother Cornelius, were by then associated with their father in the Hereford business, that there could be found on the Gudgell & Simpson ranch at Edmond, Kansas, just the bull that could carry forward the Engle program. This, he told them, was a youngster named Beau Blanchard. He was, said Frank Gudgell, a son of Mousel Bros.' Beau Mischief, while his dam was Blanche 23d, a daughter of Beau Brummel and likewise of intense Anxiety 4th ancestry.

Soon after that Denver show Jesse Engle visited the ranch at Edmond and looked over the bulls that were for sale. He liked some of the others about as well as the one that had been cited, but decided to follow Frank Gudgell's suggestion as to Beau Blanchard's potential. He never regretted his decision. The price paid for the bull, then just a few months past a year old, was $200.

A year later Jesse Engle returned to the Gudgell & Simpson establishment and purchased a number of daughters of such noted bulls as Domino, Beau President, Bright Donald, Beau Picture and others, selecting some of the very cream of the herd. These, together with the female descendants of the original 1901 purchases, provided Beau Blanchard with a formidable harem, and from that point forward for a decade and a half the Engle name loomed brightly on the Hereford horizon. And under Engle guidance, Beau Blanchard came to be generally considered the greatest breeding son of Beau Mischief, and one of the most noteworthy sires of the Anxiety 4th family.

Jesse Engle, having found in Beau Blanchard a herd bull of the breeding and type he liked, thenceforth pursued a program of linebreeding to that sire to a fairly intense degree. When the uniformity of the Engle show strings was commended by fellow breeders, Engle was quick to point out that most of the animals were the product of the herd's linebreeding program. "We attribute much of their uniformity of type and character to this fact," he once wrote. And added: "The more intense the breeding the more uniform the type."

But being the practical cattleman that he was, he emphasized again and again that no such program could hope to succeed without a herd bull of the right sort, and linebred from ancestry of proven prepotency. And it would be folly, he went on, to attempt to intensify certain bloodlines without equally emphasizing selection. Such an effort, he added, would result in "breeding pedigrees only."

The results achieved attested to the soundness of the Engle program. From a cow herd that seldom numbered more than 75 head, the Engle representatives held their own in major competition against strings sent out from much larger herds, and breeding stock went from the Engle farm to important herds across the country.

After exhibiting more or less locally at smaller fairs, beginning in 1909, the Engle herd first made its presence felt on the national scene at the 1916 American Royal when Belle Blanchard, a daughter of Beau Blanchard, was declared junior champion female. Other significant winnings continued at the Royal and other leading fairs and expositions during the subsequent 10 years as Engle show strings ranged across the country all the way from the Rocky Mountains to the East Coast with results that gained recognition for the Beau Blanchards as one of the country's foremost Hereford families.

At Denver's 1918 National Western Stock Show, for example, where the Engle herd was represented by only five head, the winnings included first on senior bull calf, second on junior yearling heifer, first on senior heifer calf, first on junior heifer calf, second on young herd, first on calf herd, first on get of sire and third on produce of dam. At the International Live Stock Exposition in Chicago the following year, two daughters of Beau Blanchard were the senior champion and junior and grand champion females, while a Beau Blanchard foursome ranked second in the get-of-sire lineup.

Such performance naturally led many breeders to buy Beau Blanchard-bred herd-bull prospects or to add Beau Blanchard daughters to their breeding herds. Some did both.

When the herd was dispersed in 1927 following Jesse Engle's death the previous year, the top-selling bull went to T. E. Mitchell & Son of Albert, N. M., in which at least two Beau Blanchard bulls had been in previous service.

A leading buyer of females was Dr. C. H. Harris of Harrisdale Farms at Fort Worth, Texas, who selected a dozen females, including five daughters of Beau Blanchard for a herd that was destined also to become a history-maker in its own right. *The Hereford Journal*'s headline on its report of that auction in which 104 head were sold was: "Engle Dispersion Greatest Hereford Sale in Recent Years."

In addition to the Beau Blanchard influence contributed to the industry through his sons, some daughters and granddaughters also left their mark on the breed. A few of his daughters were sent to Colorado by the Engles in 1923 to be mated with Fulscher & Kepler's Prince Domino, whose fame was then at its zenith. One of them was the American Royal blue-ribbon winner, Belle Blanchard 61st. The resulting calf, named Prince Domino 9th, went in 1925 to the Caerleon Ranch of W. A. Crawford-Frost of Nanton, Alberta, where he founded a family which exerted a powerful influence on Canadian Herefords for many years.

Miss Blanchard, a daughter of Paul Hahnewald's Beau Blanchard 5th by Beau Blanchard, from a mating with Dandy Domino 2d by Prince Domino at Banning-Lewis Ranches, Colorado Springs, Colorado, produced Prince Domino 101st which not only was acclaimed champion bull of the celebrated Golden Anniversary Hereford Show at Kansas City in 1932 but went on from there to become the founder of the great Colorado Domino family.

An extraordinary performance by another female of the Beau Blanchard line came when Bessie Ann Blanchard, daughter of C. A. Meyer & Sons' Beau Blanchard 94th by Beau Blanchard, was senior and grand champion female at the American Royal and International shows for two successive years, 1928 and 1929, the first time this feat had been accomplished at these Register of Merit events, and rarely duplicated since then.

Since most of the Beau Blanchards are recognizable by the family name, perhaps special reference should be made to the Beau Baldwins, which were descended at Baldwin Ranch at Pleasanton, California, from Beau Blanchard 155th, a grandson of Beau Blanchard and blue-ribbon winner as a junior calf at the 1923 American Royal before topping that year's Royal auction sale at $3,000 in going to the California herd. The Beau Baldwins gained considerable popularity in numerous herds during following years, among them being Beau Baldwin 32d by "the 155th" which came back to Missouri in 1931 to win the Royal bull show's senior and grand championship for Baldwin Ranch.

So it is, that as this is being written—some 60 years after Beau Blanchard's death—the family's influence continues to be perpetuated in various herds to which his more immediate descendants contributed their prepotent inheritance during their productive lifetimes.

# OTTO FULSCHER

HOLYOKE, COLORADO
1875-1967

OTTO FULSCHER long ago came to be recognized throughout the Hereford world as one of the master improvers of the breed. Emigrating at the age of five from his native Germany, his family lived in Missouri for his first five or six years in the United States, then went on to Colorado where the father secured land in the Holyoke neighborhood. The little town in the northeastern corner of the Centennial State, in due course, became one of the most familiar addresses on the Hereford map as a result of Otto Fulscher's activities during a period of well over a half-century.

This man who did indeed become a Hereford-industry legend in his own time began his career in the livestock field when only a mere lad by trading his shotgun and five of the six silver dollars which constituted his total cash resources at the time for a grade Shorthorn heifer. Before he got her home he made an even trade with a neighbor for a steer, which he sold a few weeks later to a country cattle buyer for $25. Thus was the beef-cattle career of one of America's greatest registered Hereford breeders launched.

A few years later while engaged in buying and selling cattle, and running a small grade herd of his own, Otto Fulscher met John Reimers, a prominent cattle feeder of Grand Island, Nebraska, who persuaded the still-young Coloradoan to go to work for him as a cattle buyer. Fulscher once related what a profound effect Reimers had upon his future career as a beef-cattle breeder in these words: "He taught me what to look for in selecting, what to breed for in my own operations, and how to care for and feed my cattle." Fulscher also long remembered his early employer's admonition when he went on the road seeking feedlot replacements to "let them alone if they are not the kind that will make choice beef."

Although he admired good registered Herefords, Fulscher considered himself strictly a commercial operator through most of the first decade of the 20th century. It was by accident that he made his first acquisition of purebreds in 1909. While on a buying trip in Nebraska he was told by a man with whom he was trying to deal that a nearby neighbor had six purebred heifers for sale. Without a great deal of enthusiasm, Fulscher went to see them, and the upshot of it was that he bought the bunch for $60 a head. Although he was unaware of it at the time, those heifers traced back to a bull named Anxiety 4th, of which he probably had never heard. They were destined to comprise the nucleus of the future Fulscher registered Hereford herd.

Through the influence of those heifers he began to attend the annual National Western Stock Shows in Denver, which whetted his interest still further. On one or two occasions he tried to buy a bull there, but not until the 1913 show did he see one which stopped him in his tracks. This eight-months-old calf exactly struck his fancy.

Upon inquiry, he learned that the youngster's name was Beau Aster, that he was a son of the Gudgell & Simpson-bred Beau Mischief, and that his exhibitors, Mousel Bros. of Cambridge, Nebraska, would not put a price on him because he had been listed for an auction to be held the following month at Grand Island, Nebraska. Robert D. Mousel, a member of the Nebraska firm, visited so persuasively with the new man that a decision was made by the latter to go to the sale with the idea that the calf could be bought for $300. As events turned out, it required a Fulscher bid of $600 to get Beau Aster, but that purchase was a vital element in laying the foundation for much that Fulscher was to accomplish in subsequent years.

When the first daughters of Beau Aster began to reach the age of around one year, their owner started to anticipate the need for another bull. By then thoroughly aware of Gudgell & Simpson's contribution to breed progress, he inquired as to the availability of suitable prospective sires of serviceable age at the Missouri sale headquarters of this noted firm. Two or three were highly recommended, and Fulscher in the spring of 1915 embarked on a motor trip with the intended destination of Independence, Missouri, for the purpose of buying a bull.

On the way he stopped at the Mousel Bros. ranch for a little visit, and

*With a then-recent issue of* The American Hereford Journal *in hand, Otto Fulscher looked up as the photographer aimed the camera for this picture taken at the Fulscher family home.*—Hereford Journal *photo.*

persuaded Robert Mousel to accompany him to Edmond, Kansas, some 50 miles south of Cambridge, where the bulk of the G. & S. breeding herd was maintained. The object was to inspect the dam of one of the bulls which he was planning to consider at Independence. At Edmond the visitors happened to find Charles Gudgell, and Fulscher told him of the purpose of their visit, stressing that he did not like to make a bull-buying decision without seeing the dam, and possibly the granddam and other members of the family, of the animal in question.

Fulscher later recalled that dozens of magnificent brood matrons were viewed on that eventful day, April 6, 1915, including such great eye-filling history-makers as Donna Anna 22d, dam of the G. & S. herd bull, Domino, and Lady Stanway 9th, which numbered among her sons, Bright Stanway, then rated as the top Gudgell & Simpson herd bull. When time came for the noon-day meal, the visitors joined the ranch crew in the bunk-house dining hall for dinner and talking on a variety of topics. The subject of hay stacking came up, and Gudgell spoke about an ingenious hay-stacking device which had originated in his mind, and in which he took considerable pride.

After dinner, Gudgell suggested that his visitors might like to see the stacker, so he led the way toward a nearby field where it stood. Some ranch buildings and corrals were along the route, and seeing some calves in one of the corrals Fulscher lagged behind the others to have a look. As he looked beyond the fence he saw a calf that seemed to his mind's eye to "have everything." He proceeded into the corral and prodded the calf to get up, and the more he saw of the youngster the more impressed he was.

"The calf filled my eye," Fulscher related many years afterwards. Mousel confirmed the Colorado man's judgment, so the veteran G. & S. herdsman, William (Bill) Hendry, then was asked as to the identity of the six-months-old youngster's dam. He told the visitors that she was Lady Stanway 9th—the impressive cow which had been seen earlier in the day and cited as the dam of Bright Stanway. Although Fulscher was in the market for a bull of serviceable age, this calf seemed to be of such extraordinary promise that the Colorado man asked that a price be quoted and Gudgell said: "$500."

At Gudgell's insistence, the two men, Fulscher and Mousel, decided to go on to Independence, and while the bull they especially wished to see, along with a couple of others, was impressive, Fulscher simply could not deny the calf at Edmond, on which he had been given an option. So he told Charles Gudgell's son, Frank, who was on the job at Independence, that he would exercise the option, and left shipping instructions. Since the calf had not yet been recorded, Fulscher requested that his name be registered as Prince

*This is Prince Domino as he looked during the latter stages of his remarkable career as perhaps the Hereford breed's most influential sire of the 20th century. It may not be as glamorous as earlier pictures of the bull but it emphasizes the masculinity and character, even at an advanced age, which were Prince Domino trademarks.—Hollard photo.*

*The Fulscher ranch, 16 miles northeast of Holyoke, was the unpretentious home of Prince Domino and his predecessor, Beau Aster, as well as their successors, the Real Prince Dominos. Beyond this set of brood cows, at extreme right, is the sale barn which was the setting of several notable auction events.—Smith photo.*

Domino. The calf's sire was Domino, whose dam was the great Donna Anna 22d, while his own dam, as previously noted, was the dam also of the famous Bright Stanway, then the chief G. & S. herd bull. How much more powerful could the maternal influence have been?

It is no exaggeration to state that this incidental stroll toward a hay field, along with Fulscher's curiosity and inquisitiveness, and the resulting purchase, marked a major milestone—perhaps *the* major turning point—in American Hereford history of the 20th century. Prince Domino was used in the herd only lightly during the summer and fall of 1915, another older bull having been acquired for temporary service. When the new bull's first calves began to arrive in the fall of 1916, the unemotional Fulscher could scarcely contain his elation. They were all he had hoped for, and then some. And visitors to the ranch eagerly confirmed the owner's judgment.

While Fulscher obviously bought Prince Domino with the aim of mating him with the daughters of Beau Aster, no more than a few of the latter's daughters could have made their appearance in the Fulscher herd when Prince Domino's selection was made. Beau Aster at that time was not yet three years old, and his oldest calves had just passed their first birthday. Nevertheless, if the Holyoke breeder had had a herd of mature daughters of Beau Aster before his eyes, and had selected Prince Domino with the express purpose of mating the bull with them, he could not have made a happier choice. Under the banner of Fulscher & Kepler, the Prince Domino-Beau Aster nick achieved international renown as one of the most successful in the history of American Herefords. It is to Otto Fulscher's credit that the products of this mating established Prince Domino's reputation as a sire.

By 1918 the Prince Domino calves began to compete in the Denver show, and at the 1919 event a son and a daughter were made the champions of the Western Hereford Futurity, then a feature of the big Rocky Mountain event. Both were from Beau Aster dams, and the bull, Leo Domino, eventually was bought by Banning-Lewis Ranches of Colorado Springs where he was re-named Dominator, and proceeded during subsequent years to become one of that firm's key foundation sires.

A year later Fulscher & Kepler exhibited another sensational string of Prince Domino calves, one of which became champion female of the great Denver event. She, too, was from a Beau Aster dam. Princeps Domino, top bull calf in that string, brought $7,300 in the Denver auction sale from J. N. Camden of Versailles, Kentucky, who exhibited him to the junior and grand championship of that year's Chicago International. The seven Prince Domino youngsters which Fulscher & Kepler exhibited at that 1920 Denver show were sold in the auction for $21,225, and it was reported that an offer of $51,000 for Prince Domino himself was refused.

Reference to Princeps Domino brings up a point which Otto Fulscher once emphasized. This calf's dam was not, he pointed out, a Beau Aster cow. He went on to state that while there was a general impression that Prince Domino's reputation should perhaps be shared with Beau Aster, the fact was that comparable success was recorded when the new bull was mated with cows of other families, as in the instance of Princeps Domino, whose dam was by Princeps 20th, a grandson of the early Mousel bull, Princeps 4th. And a study of breed history records numerous other examples, including such instances of blood intensification as exemplified in Real Prince Domino, sired by Prince Domino and out of a daughter of Prince Domino, and Real Prince Domino 33d, sired by Real Prince Domino and out of another daughter of Prince Domino. Both of these bulls became history-makers in their own right, and are to the credit of Otto Fulscher's imagination and daring.

Fulscher made another move which contributed to the expansion of Prince Domino's reputation when in 1924 he sold a half-interest in the bull to Ken-Caryl Ranch of Littleton, Colorado. This transaction was a master stroke in enabling Prince Domino to achieve new heights of national acclaim because Ken-Caryl was a regular exhibitor at most of the country's leading shows. This step enabled Prince Domino to rise to leadership in the original Hereford Register of Merit, a place he held uninterruptedly from 1927 until 1948 when he was replaced by a descendant. As late as 1957, his name still ranked fifth in that Register even though by then he had been dead for 27 years. And it is of further interest to note that three of the four bulls that ranked above him in 1957 were his direct descendants through their sires, and that the fourth one, TR Zato Heir, was descended from Prince Domino through his dam.

For around a half-dozen years, beginning in 1928, Fulscher was associated in an advisory capacity with Wyoming Hereford Ranch at Cheyenne. The breeding herd of Fulscher & Kepler at that time was largely consolidated with that of WHR, and approximately one-half of Prince Domino's time was shared between Ken-Caryl and WHR until the old bull's death in 1930 at WHR where a monument, erected in recognition of his greatness, still stands. However, there was no change in his ownership status, which continued to the end to be solely Fulscher & Kepler and Ken-Caryl.

After the arrangement with WHR ended around 1934, Fulscher re-entered the Hereford business on his own, his association with his original partner, Leon Kepler, also having ended some time earlier. The Prince Domino influence continued to be the paramount factor in his operations right to his life's end, in 1967, at the age of 91. His son, Max Fulscher, a partner in his latter years, continued the herd's operation thereafter, his son, Gary, eventually joining him in the operation of Fulscher Herefords.

A recital of the history-making Herefords attributable to Prince Domino would be voluminous and repetitious. The extent to which his sire-line descendants continue to this day to dominate to breed's ancestry is evident in the fact that their names occupy 78 of the 107 pages in the current volume of *Hereford Lineage.* Undoubtedly many of the others reflect additional Prince Domino influence through their dams' inheritance.

Thus evident is the extent to which the Hereford breed yet today is indebted to Otto Fulscher. Except for his judgment and the influence of his constructive effort during a period of well over a half-century, the course of the breed's history in America obviously would have been far different.

# Dan D. Casement

MANHATTAN, KANSAS
1868-1953

ONE of the most remarkable men ever to earn his spurs in the Hereford field surely was Dan D. Casement of Juniata Farm, Manhattan, Kansas. Here was a man born, it might appropriately be said, with a silver spoon in his mouth. An Ohioan by birth, he was graduated from Princeton University, and then earned his Master's degree in law at Columbia University in New York City. Almost immediately thereafter came a decision which eventually made him one of the most distinguished and distinctive figures in the entire livestock industry.

With no particular background in the field, and no formal training in agriculture and livestock production, he cast his lot with the land. And with Herefords. In the early 1890's, when just out of college, he engaged in a ranching partnership with a college roommate on land their fathers had acquired below Grand Junction, Colorado. Here, he once related, he learned some hard lessons. But he persevered. And he succeeded.

Except for an interval of six years just after the turn of the century when he was associated with his father, General Jack Casement, in building a railroad in Costa Rica, he devoted the remainder of his 85-year lifespan to livestock, and particularly Herefords. His glory years with the Whitefaces began when he moved from Colorado in 1915 to make his home permanently in

*Hereford steers in his Juniata Farm feedlot responded wonderfully well to Casement's ministration. He delighted in lighter moments, as he once expressed it, in "currying favor" with the animals. Curry comb in hand, he always found his proteges receptive as he walked among them.—Gene Guerrant photo.*

Manhattan, although he had owned the one-section Juniata Farm, in a bend of the Big Blue River north of that town, since his 21st birthday when his father gave it to him. The General had come into its possession when he and his brother—the heralded Casement brothers of railroad-construction fame—were building the roadbed and laying track for what then was the Kansas Pacific Railroad, forerunner of the Union Pacific line of later times.

Dan Casement's attachment to Herefords was of long standing. He once wrote that the first registered Hereford bull he ever saw was on Juniata Farm when he was 16 years old. His father, of course, was the owner. From this modest first exposure to the Hereford influence, he went on to become one of the breed's most successful commercial operators and perhaps its most eloquent spokesman. Nor was his success attained through the planning and activities of surrogates. He was right out there "in the muck," as he once expressed it, working with the farm's cattle and other livestock in all seasons and in all kinds of weather. Because, he frankly pointed out, and contrary to the impression of some outsiders, his operations had to pay their way; otherwise, he said, he could not afford to keep them going.

Although he operated a modest-size herd of registered Herefords for many years at Juniata, and on some 2,000 acres of Flint Hills pastureland a few miles north of his headquarters, it was Casement's commercial Hereford operations which earned lasting fame for him. His commercial breeding herd, usually mated with bulls from his own purebred herd, yielded the superb carloads of feeder calves which earned the frequent accolades of the beef-cattle industry through their winnings from coast to coast.

But he was equally successful as a feeder and exhibitor of the carloads of finished beeves which likewise brought honor and distinction to him and to the breed in the country's best fat-stock shows. Some of these were of his own production, especially in his later years, but during the course of his unique career he fed and fitted with remarkable success representatives of several of the West's best known commercial ranches, all noted for their high quality.

He once related that in the course of picking show prospects for this carlot competition he had, as he put it, "prowled" across such famous ranges as the Crosselle, the WS, the CS and the Bells, all in New Mexico; the Matador, the SMS, the Mill Iron, Gillett's and W. B. Mitchell's in Texas; the Baca Grant and T-Bone in Colorado; Huidekoper's American Ranch in Montana, the XI in southwest Kansas, and several others similarly prominent.

One thing all of these outfits had in common was that they featured the production of top-quality commercial Herefords through the use of high-level Hereford sires. No other kind interested Dan Casement.

His devotion to Herefords on his own account dated back to his early experience on the western Colorado range. He bought his first purebred Hereford bulls at Denver in 1904. After witnessing their performance in the rugged Unaweep canyon region near the Colorado-Utah boundary, he ever afterward was an ardent champion of the breed, and a vigorous spokesman for Herefords whenever any matter of breed preference arose. His clarion battle cry in the heyday of his career in exhibiting carloads of finished Hereford beeves was "Beat those blacks," and, whether the victor or the vanquished, he never gave up his quest for the ultimate honor.

So it was as a feeder and exhibitor of fat and feeder cattle in carload lots that Dan Casement not only achieved fame in the livestock industry but also gained a spectacular kind of stature in the eyes of the general public. It became the opinion of most of his well-informed contemporaries that no man of the soil or the grasslands, before him or afterward, ever came close to matching him in that respect.

He made his American Royal debut as a carlot exhibitor in 1907. His carloads of fat Herefords first appeared at the Chicago International in 1912 and at the Denver stock show in 1913. During his career as an exhibitor, which spanned 45 years, he showed winning carloads of Hereford steers all the way from Baltimore in the east to Portland and Los Angeles on the West Coast, and at Fort Worth to the south. In 1945 an inventory of the show ribbons still in his possession, garnered in America's top-flight carlot competition, totaled

*Dan Casement as he appeared long ago in the carlot feeder cattle pens at Denver. The date is unknown but is thought likely to have been in the 1920's.*—Hereford Journal *photo.*

316, and, he added, "some have doubtless been lost and others given away."

He continued to exhibit until his final blue and purple ribbons were won at the American Royal in 1952–in his 85th year–when it was conservatively estimated that his entries had accounted for at least 350 ribbons during his long and brilliant career. The carload of feeder calves which won his final Kansas City victory were of his own breeding, as were most of the winning carloads of his later years.

Although customarily seen in the battered garb of a working cattleman while engaged with the daily tasks at Juniata, he took delight in wearing the somewhat flamboyant and picturesque attire which became his virtual trademark when at the stock shows. With his spectacles attached to a wide black ribbon and his pipe almost ever-present, he was the picture of a legend when he donned his suit of black and white shepherd's check, his brilliant weskit of Stuart plaid, and his red tie. "Thus accoutered," as he once wrote, he loved to "sally forth" on his annual pilgrimages with a determined purpose of adding still further to his prize-ribbon collection.

It is a tribute to the Hereford breed that it could attract and retain the active interest and loyalty of a man whose mental capacity and opportunities made him one of its most effective champions. He was that rare individual who possessed both conviction and the ability to express it. He could paint with words as a talented artist paints with a brush. Or, if the occasion required, he could express himself with the punch of a Dempsey, with a rare selection of pungent expletives to fit the occasion. It was once said of him that "whoever argues with him will be informed if not convinced."

He gloried in conflict in the interest of an ideal. That was the instinct which led him to fight unrelentingly the encroachment of controls and regimentation during the days of the so-called New Deal, a position from which he never retreated even for an instant.

His Hereford friends, too, always knew where he stood. He could not be a pussyfooter. He believed, for example, that the over-fitting of bulls for the showring was ridiculous, and said so. He deplored insistence upon what he termed "fancy

*Dan Casement was a "salty" and down-to-earth cattleman, yet exceptionally literate and articulate. Here he sat in the library of his home in Manhattan, Kansas, with Nicodemus, one of his favorites.–Gene Guerrant photo.*

*Dan Casement's figure was a familiar one in cattle showrings across the country for almost a half-century. Although he exhibited only in the carlot steer divisions, he usually followed the judging of the individual breeding classes. Resplendent in his suit of black and white shepherd's check and his weskit of Stuart plaid, he was pictured here in the judging arena of the 1946 American Royal show while visiting with Don Ornduff, then editor of the* Hereford Journal.—Hereford Journal *photo.*

points," which he believed had little bearing upon economic usefulness. In brief, he believed that Herefords should be bred and fed to produce choice beef with maximum efficiency and economy. In later times, this view came largely to prevail throughout the industry, and by his early advocacy Casement stood forth as a benefactor of the breed.

Many honors came to Dan Casement. He was feted by the American Hereford Association in Kansas City in the fall of 1952 in token of his life-long contribution to the advancement of Herefords and American animal husbandry in general.

He served as a member of the board of directors of the International Live Stock Exposition, and was guest of honor at a luncheon during this Chicago event in 1952 when he was praised for his dedication and efforts in behalf of the livestock industry.

He was a life-time honorary member of the American National Cattlemen's Association and a long-time member of its executive committee.

But the honor he probably prized most of all came in 1939 when his portrait was unveiled in the celebrated gallery of livestock-industry leaders at the Saddle and Sirloin Club which then had quarters in the Stock Yard Inn, adjacent to the Chicago stock yards. The citation on that memorable occasion referred to him as "Stockman, Farmer, Patriot, Exponent of the Pioneer Self-Reliance, Personal Freedom and Individual Independence That Makes the American Tradition."

How fortunate it was that a man of such caliber cast his lot with Herefords, and contributed so greatly in keeping the breed in the forefront of the beef-cattle procession!

# Albert K. Mitchell

ALBERT, NEW MEXICO
1895-1980

Few men in the modern history of America's beef-cattle industry, if ever, contributed as widely and wisely to its advancement as Albert K. Mitchell of Albert, New Mexico. Nor were his accomplishments confined merely to the Hereford field, although this was the focus of his own immediate effort. Rather, this broad-gauged man answered the call dozens of times to lend counsel and guidance toward the solution of a wide spectrum of problems touching upon agriculture and livestock production in general. It is likely that no other man in the entire field ever applied his talents and imagination more effectively in stamping his constructive imprint upon the industry.

Albert Mitchell was born into the ranching business, and his home never was other than the Tequesquite Ranch, which was established by his father, Thomas Edward Mitchell. The latter, who was born at Fairplay, Colorado, during the mining boom there, early in life became a cowboy for the Dubuque Cattle Company, an Iowa syndicate, which operated a huge open-range spread in the Tequesquite Valley of northeastern New Mexico.

So steadfast was he in the performance of his duties that in 1881 he became general manager of this outfit, and continued to run it until barbed wire and windmills sealed the doom of the open range, and homesteading became the order of the day. The elder Mitchell then staked out a homestead along

Tequesquite Creek in 1894, while still working for the Bar T Cross outfit, as the Dubuque company was known, and several other Dubuque cowhands did likewise. Most of these latter properties later were acquired by T. E. Mitchell, who also added further to his holdings by the acquisition of other Dubuque acreage.

In 1895, T. E. Mitchell and his wife, the daughter of Mr. and Mrs. Andrew Knell, took their son, Albert K. (for Knell) Mitchell, then only a few months old, to their new rock and adobe house in the Tequesquite Valley. This ranch was the home ever afterward of the man known to Hereford posterity for three generations of remarkable leadership. Thus, from this rather remote corner of New Mexico, some 85 miles north of Tucumcari and 18 miles northeast of Mosquero, Albert Mitchell made his presence strongly felt for far more than a half-century not only in the cattle business but also in other ways across America's expanse.

It is related that the young Albert required no urging to learn the cattle business from the ground up, and from that beginning he assumed new responsibilities step by step. It is ranch tradition that he could handle a rope when four years old, and that at eight was more expert than some of the ranch hands. Standing on the roof of the stone cellar near the house he learned well the art of dropping the noose over the tops of nearby fence posts. Albert's roping skills continued to be evident 60 to 70 years later as he regularly performed this task at branding time on the Tequesquite range. And continued to do it with a proficiency which aroused the admiration of seasoned onlookers, such as Dan Casement, noted cowman of Manhattan, Kansas.

Speaking in 1949 on the occasion of the presentation of Albert Mitchell's portrait for hanging in the celebrated gallery of livestock industry figures at the famous Saddle and Sirloin Club, Casement said: "It is my considered judgment that Albert Mitchell is today the greatest cowman in America, and that he has amply proved his right to that distinction in all departments of our business."

After citing the honoree's "intelligent management of the large herds of registered and commercial Herefords" on the Tequesquite and "his forceful and successful handling" of the nearby Bell Ranch, a vast property which he managed additionally for some 16 years, Casement went on:

"Great cowman that he is, Albert Mitchell is an equally great cowhand. No one will dispute that he will do to ride the river with. No bronc is too salty for him to break; no renegade too rough for him to rope regardless of the nature of the terrain. To enumerate the jack-pots he has survived and the bones he has broken would require a big tally sheet. . . .

*Albert Mitchell was equally "at home" astride a Tequesquite mount or in the cockpit of a private airplane. He began to fly his own plane in the early 1930's, and no one can say with exactness how many tens of thousands of miles he traveled during the next 45 years and more.—A Tequesquite photo.*

"On the bunch ground he is always where the dust is thickest; at the weaning pens he is always in the gate, looking in the eye the frantic cows as they come out a-stoopin'. The hardest and most dangerous tasks he invariably delegates to himself. Albert doesn't take his dallies. He's a tie-fast man. Whenever he shakes out a loop and drops it on a critter it is his or he's its."

Now to backtrack a bit. When Albert's father, T. E. Mitchell, entered the cattle business "on his own," it was with commercial Herefords, using purebred bulls. In the late summer of 1896, the senior Mitchell once related, while on a trip to Channing, Texas, to buy some bulls from William Powell, it occurred to him that this was the proper time to execute a commission given to him by his father-in-law, Andrew Knell, whose ranch was near the Mitchell layout, to buy for Knell a few select heifer calves.

He continued: "Thus were shipped to Texline, Texas [nearest delivery point to the ranch], in the same car with the bulls but separated from them by a partition, six of the best heifer calves I could find among the year's output of the Powell ranch. I presented the calves to Mr. Knell."

Shortly afterward, T. E. Mitchell contracted for the bull calves from these heifers, this arrangement continuing until 1913 when Mr. Knell's advancing years made it necessary for him to retire. T. E. Mitchell stated in an article in *The American Hereford Journal* that he and his son Albert at that time purchased 158 females from Mr. Knell, and that these descendants of the William Powell heifers constituted their registered Hereford foundation herd, to which only one outside cow was added in later years. This was the Gudgell & Simpson-bred Bright Duchess 67th, for which Albert Mitchell bid $5,950, one of the top female prices of the day, in the record 1920 auction sale of Mousel Bros. of Cambridge, Nebraska. This daughter of Beau President had an attractive bull calf at side by Choice Stanway and was bred back to Mousel Mischief, a son of the famous Beau Mischief.

Now another look back. After early schooling at home by way of a governess, since the Mitchell ranch was remote from any school, the youthful Albert was induced somewhat reluctantly to attend a preparatory school at Pasadena, California, and then Occidental College in Los Angeles. In 1915 he entered Cornell University at Ithaca, New York, to major in animal husbandry. He was graduated in 1917 with a Bachelor of Science degree, returned home briefly, and then did a stint of military duty. When it was over, late in that same year, he returned to the ranch to stay, finding upon his arrival that his father had a present for him. Into the 23-year-old Albert's lap he dumped the management and operation of the entire ranch and its Herefords.

Having learned early in life how to accept responsibility, Albert grasped the opportunity and moved forward aggressively. Less than a year later he attended the annual auction of Wallace and E. G. Good at Grandview, Missouri, and bid the top price of $6,500 to get the 1,900-pound blue-ribbon senior yearling Good Donald 3d. The performance of this impressive young bull, whose sire was bred by Kentucky's W. H. Curtice and whose dam was the daughter of a bull bred by Mousel Bros., led to the purchase of two other Good Donald bulls from the Goods, and this line thus became an early contributor to the Mitchell herd's progress.

In *The American Hereford Journal*'s first Herd Bull Edition, issued in 1923, the two-page advertisement of T. E. Mitchell & Son listed a sire battery of 14 herd bulls, mostly descended from Gudgell & Simpson lines, and all selected, as was once pointed out, on the basis of "size, bone and quality." The last one on that list was Mischief 5th, a heavy-boned, year-old grandson of Bright Stanway and Beau Mischief, for which Albert Mitchell bid a modest $1,250 in the 1921 sale of Mousel Bros. This bull became the progenitor of a noted line of Mischiefs which went out from the Tequesquite Ranch to stamp their imprint, during the middle years of the 20th century, on generations of good Herefords in many leading purebred herds. But the primary market then and afterwards was the range bull trade, with commercial ranchers across the Southwest returning again and again to Mitchells to buy reputation range bulls in quantity.

A few years later, in January, 1930, Albert Mitchell selected another herd-bull prospect which forever stamped upon the Hereford consciousness Tequesquite Ranch as the source of the Mischiefs and the Huskers. The latter line was descended from Husker Mischief, Mousel-bred son of Advance

Mischief by Prince Domino Mischief by Prince Domino, and out of a granddaughter of Beau Mischief. He had been undefeated senior yearling bull at the leading shows of the 1929-30 season. There was a time in breed history not so long ago when the Mischiefs and Huskers were unsurpassed in prestige throughout the Hereford industry, and hundreds of them served both in leading registered and commercial Hereford herds across the country.

The Mitchell registered herd grew rapidly after its beginning. By the summer of 1919, a Mitchell advertisement in *The Hereford Journal* indicated that there were 383 cows and heifers of breeding age in the herd. Two years later, in the 1921 fiscal year, T. E. Mitchell & Son recorded 638 calves in the American Hereford Association, leading all other herds in that respect. In the years which followed, the Tequesquite Ranch has ranked either at the top or among the leaders in the number of calves registered annually. Although no exact check has been made, it is entirely possible that in the 60 years between 1920 and 1980, the Mitchell firm recorded more Herefords than any other in this country.

All the while he was heavily engaged in running both the Tequesquite and Bell Ranch operations, Albert Mitchell was accepting heavy responsibilities in organizations involving the Hereford business and the beef industry in general. One of the first of these was the presidency of the American Hereford Association, to which he was elected in 1929, when only 35 years old. In a time when disharmony briefly prevailed in the industry, Association members again turned to Albert Mitchell in 1956 in confidence that his steadying and forward-looking influence would prevail, as it did. He served on the Association's board of directors for several years and no one can say how many times his unofficial counsel was sought by those holding official posts. But it probably is no exaggeration to state that his influence prevailed over a longer period than that of any other breeder in the organization's history.

Albert Mitchell, as if to affirm the point that it takes a busy man to get things done, served with equal dedication in executive capacities in numerous other organizations. He was president of the New Mexico Hereford Association and the New Mexico Cattle Growers Association, and was named the latter group's "Cattleman of the Year" in 1963. Twice he was named to the presidency of the American National Live Stock Association– in 1936 and 1937. This is the organization now known as the National Cattlemen's Association. He formerly headed the National Live Stock and Meat Board, was president of the International Live Stock Exposition, and later was named honorary chairman of the board of the Chicago show. He held top posts with the National Western Stock Show Association, and also served as judge in the carlot bull division of the big Denver event. Not only

*One of the Tequesquite herd bulls was in the front rank as two cowboys move some of the cows and their calves toward a new location. Herefords of the caliber of these spread the fame of the Mitchell and Tequesquite names from coast to coast and beyond.* —*Caplin photo.*

did he serve as president of the American Quarter Horse Association, but was one of the group that was instrumental in bringing unity to producers of this breed in forming the association.

He also rendered leadership in other services to agriculture and livestock organizations, such as the National Live Stock Tax Committee, the Inter-American Conference of Cattlemen, the President's National Agricultural Advisory Committee, and the Foundation for American Agriculture. He served as chairman of the executive committee of the Lovelace Foundation for Medical Research, as a member of the Board of Regents of New Mexico State University, and of the Board of Trustees of Cornell University, his alma mater. He also was a director of the New Mexico Boys' Ranch, and succeeded his father as a director of the Regional Agricultural Credit Corporation of Santa Fe, New Mexico, serving from 1934 through 1938. He was a member of the original Board of Directors of the statewide Albuquerque Production Credit Association, continuing to serve this organization for many years, and also held several other significant financial posts.

It scarcely is possible to cite all of Albert Mitchell's achievements without missing some but another important post he held was that of chairman of the Board of Trustees of the National Cowboy Hall of Fame and Western Heritage Center at Oklahoma City, of which he later became honorary life chairman of the board. In 1976 he was named to the Cowboy institution's Hall of Great Westerners. It is of interest that his son, Albert J. Mitchell, is a member of the Board of Directors of the Cowboy Hall as this is written, thus following in his father's footsteps. Likewise, the son several years ago assumed the active management of Tequesquite Ranch operations, just as T. E.

Mitchell was succeeded by his son, Albert K. Mitchell. Another parallel that may be drawn is that Albert J. Mitchell was elected president of the New Mexico Cattle Growers Association in 1976, a post his father held many years earlier.

A significant public service that must not be overlooked in relating Albert K. Mitchell's career in the cattle industry came in the late 1940's and early '50's when he served as chairman of the Livestock Industry Advisory Committee named by the U. S. Secretary of Agriculture to investigate the problem and make recommendations in connection with a severe outbreak of foot-and-mouth disease in Mexico. After a tremendous effort, the campaign succeeded in bringing the scourge under control. It was said that Chairman Mitchell's ability to speak the language enabled him to work amicably and effectively with the Mexico authorities who were directly involved in the effort.

Two great Register of Merit Hereford shows were dedicated to Albert K. Mitchell, a signal honor indeed! The first was in 1971 when the huge Register of Merit Hereford Show at the National Western paid homage to him. Then in 1974, his New Mexico neighbors chose to designate their state fair's Hereford show at Albuquerque as the Albert K. Mitchell Register of Merit Hereford Show. Few Hereford personalities have been thus honored once, much less two times.

Nor did Albert Mitchell shun legislative responsibilities in his state. He served two terms as a member of the New Mexico state legislature, represented New Mexico on the Republican National Committee for 23 years and was vice-chairman of that body for three years. He was nominated by his party both for governor of his state and for United States senator.

In 1978 he was the recipient of the first Golden Spur Award, which is presented annually to honor "the individual who has brought special distinction to the livestock and ranching industry." In thus citing Albert Mitchell, the Ranching Heritage Association at Lubbock, Texas, in cooperation with several other industry organizations, brought honor to itself.

One more point: how did one man cover so much territory? By airplane! Albert Mitchell began flying his own light plane in 1934, and no one can say how many thousands of miles he covered. By 1968 he had logged more than 4,000 hours at the controls, and undoubtedly many more thereafter. This, in addition to tens of thousands of miles flown in commercial planes.

Truly, Albert K. Mitchell was uniquely a man of many parts, and it was to the Hereford breed's great advantage that a man of such remarkable talent and energy devoted a lengthy lifetime, which ended only in 1980, to its advancement.

# R. J. Kinzer

KANSAS CITY, MISSOURI
1876-1952

A MAN closely associated with R. J. Kinzer in the Hereford field for more than 40 years once undertook to speak in behalf of the entire industry regarding Kinzer's contribution to the breed during his tenure as secretary of the American Hereford Association. These were that speaker's words:

"Members of the Hereford Association throughout the country, together with thousands of others identified in one way or another with this great industry, would join in recognition of the efficiency of his leadership. They would honor and admire him for his ability and application, for his shrewdness and sagacity, patience and perseverance, for his resourcefulness and for his loyalty that he has so splendidly exemplified as the active head of the purebred Hereford organization."

Those who knew him best, and observed his activities at close range, certainly concurred in that appraisal. Many likewise regarded him as a genius in the exercise of his talents in advancing the interests of the Hereford breed. In explanation, it has been said that "genius is nothing more than an infinite capacity for taking pains." As a breed association secretary, and in his case this amounted practically to a personal leadership, Secretary Kinzer–if this statement be true–surely qualified as a genius.

No day too long, too cold or too hot, no trip too difficult or wearisome, to find him unwilling to finish a task he had started. No problem was too big for him to undertake a solution, no office detail too trivial to gain his close study, whether it involved a $100,000 purchase of securities for the Association's portfolio or a conflict in birth dates on a registration application–the latter something that an office clerk might well have straightened out.

The results of his efforts in behalf of the Hereford breed and Hereford breeders during an active career which spanned a third of a century stand forth as an unchallenged testimonial to the success of his efforts. To accomplish his aims he was willing to use every personal tool at his command–persuasion, logic, tact and diplomacy, jollying and even, on occasion, a bit of quiet coaxing. But it must be noted also that he was a man of few words, and when those words were unavailing he was not adverse to bringing a bit of well-directed pressure to bear. It was generally agreed that his role throughout most of his years at the operational helm of the Association was a decisive one.

To go back to the Kinzer beginning, he was born and reared on the family farm at Bangor, Iowa, in the central part of the Hawkeye State. In due course, he was graduated from the institution that later became known as Iowa State University at Ames with a degree in animal husbandry. While a student there he won first in the collegiate livestock judging contest at the International Live Stock Exposition in Chicago, and on his graduation in 1901 he joined the animal husbandry staff at the college. A year or two later he went to the animal husbandry department at Kansas State Agricultural College at Manhattan, Kansas, which later became Kansas State University. He stayed there for eight years, and was head of the animal husbandry department when he accepted the position as secretary of the Hereford Association, effective on January 1, 1911.

Obviously he had established an outstanding record in the Kansas post, which is what drew the attention of the Hereford Association's directors toward him. In reporting his selection, *The American Hereford Journal* said: "He is an expert feeder and exhibitor, as has been attested many times by the success of the college entries under his charge at the various livestock shows. He is an expert judge of livestock, and his services in that line have been in demand. This fall he has served as judge at some fair every week of the fall show circuit. His interest in Hereford cattle is well known. He has always been regarded as one of the cleanest, cleverest, most honorable and likable men connected with college work. . . . He is a young man, admirably equipped for the work."

The expansion of the Hereford industry during Secretary Kinzer's tenure is highlighted by the statistics. Registrations during the year before he assumed the helm were 23,614. By 1920 the total reached was 102,503, and it went from there to just under 300,000 by the time of his retirement in 1944, when he was advanced to the position of chairman of the board which was created for him.

In his incumbency as secretary his name was affixed to a total of 3,713,458 registration certificates, and in many of his years as secretary Hereford registrations numbered more than those of all of the other beef breeds combined–in fact, in some years twice as many as the combined totals of the others. Those years saw the breed going ahead to new heights that even the most optimistic would not have dared predict.

Without a doubt, R. J. Kinzer became the best known man connected with the Hereford industry in America. Neil M. Clark, a writer for the one-time leading national agricultural magazine, *Country Gentleman*, once stated that "If a jury were selecting the most influential living cattleman, they'd have to consider him though he owns no cattle." Clark went on: "How does a man make his influence felt? Some write. Others make speeches. R. J. prefers personal contacts. He isn't much of a talker. Friends say that if something is under discussion and he 'sort of grunts' and walks away he's against it."

R. J. Kinzer was at his best in working behind the scenes, and his success at convincing others on matters having to do with breed advancement and association policies was brought about quietly through the man-to-man

*Secretary Kinzer sat at his desk as the board of directors of the American Hereford Cattle Breeders Association gathered on February 3, 1913 in his office at the headquarters of the Association, then in a downtown office building at 1012 Baltimore Avenue in Kansas City. Around the table, left to right: J. A. Shade, Kingsley, Iowa; Dr. T. F. DeWitt, Colorado Springs, Colorado; Robert H. Hazlett, El Dorado, Kansas (treasurer); Overton Harris, Harris, Missouri; Phil C. Lee, San Angelo, Texas; Kinzer, and F. C. Giltner, Eminence, Kentucky (president). Not present was another director, Warren T. McCray, Kentland, Indiana.*

method. He did his work with a superlative degree of ability yet without ostentation and with a studied desire to avoid the personal limelight.

Many a breeder during the heyday of Kinzer's career admitted gladly that it was the advice and generous help he received from the secretary which started him on the right track and helped keep him there. It would be more than interesting to know how many herds were founded during his one-third of a century as secretary on seedstock selected according to his advice and, in many instances, with his actual assistance in selection. It might perhaps be even more interesting to know how many established breeders turned to him for the same generous assistance in the solution of their problems, whatever they may have been.

Nothing appeared to be too much trouble if its accomplishment would work toward the advancement of the breed. He somehow managed to keep a finger on the pulse of the entire industry, and guided many a man toward Herefords who had not to that point quite made up his mind as to breeds. Instinct seemed to lead him to men of means who were ready to enter the beef cattle business. This consistent infusion of new interest in Herefords was a significant element in the breed's expansion and economy, and was greatly to his personal credit.

When Kinzer assumed the secretaryship, its treasury was virtually empty, and its small office force was maintained in rented quarters. The great accomplishment of his first decade was to guide the erection of a new home for the Association, an imposing marble-columned structure at 300 West 11th Street in Kansas City, which was dedicated in 1920. Years after the Association left it for still larger quarters, this impressive structure continued to be pointed out to visitors as one of the most attractive buildings in the downtown business district. When the Association staff moved in, it immediately became a beehive of industry activity such as even the man who was instrumental in its erection had no way of anticipating. The Hereford, incidentally, was the first beef breed to have its own Association headquarters building, and one of the first, if not the very first, among all kinds of livestock.

Shrewd in business matters and naturally thrifty, he gave to association business even closer scrutiny and attention than he gave to his own, and instituted conservative financial practices which, in conjunction with the breed's expansion under his direction, led to a substantial build-up in the Association's monetary reserves. The financial strength of the Hereford Association, long the envy of other livestock organizations, was guided by Secretary Kinzer, and a great portion of the more than $5,000,000 balance sheet of the post-Kinzer era was mainly attributable to his foresight.

One of his most far-reaching acts was his introduction on January 1, 1918,

*In characteristic pose as he checks a point in his catalogue, R. J. Kinzer sat at the ringside as the dispersion sale of Jack Turner's Silver Crest Herefords progressed on a June day in 1946. By then he had been named chairman of the board of the Hereford association, having relinquished the secretaryship two-years earlier. Roy Turner, owner of Turner Ranch at Sulphur, Oklahoma, sat beside him.—American Hereford Association photo.*

of a tattoo system which he devised. This immediately became a registration prerequisite, the accuracy of the records thereby becoming more carefully assured and preserved through this practically infallible system.

Roland Jacob Kinzer—to state his name fully—was uniquely connected with the two greatest Hereford exhibitions of his times. One was the great Golden Anniversary Hereford Show, staged under his direction in connection with the 1932 American Royal in Kansas City to mark the Association's 50th anniversary. A graphic illustration of the advance and expansion of the breed, especially during the two decades or so of his secretaryship, was the amazing total of 4,670 Herefords which made up this record display. Of these, 885 head were registered Herefords in the individual breeding classes. The other 3,785 were steers in individual and carlot divisions. Exhibitors numbered 112 from 22 states, with entries led out for competition in the individual breeding cattle classes numbering up to 65 head per class.

The second of these two landmark Hereford shows was staged in conjunction with the American Royal and named the "R. J. Hereford Royal," held after his retirement in recognition of a lifetime of service to the breed. Staged in 1948, four years before his death in the pasture of a Florida Hereford breeder whose herd he was looking at when fatally stricken, it also was a great event, with some 600 registered Herefords from 24 states competing in the individual breeding classes.

It is regarded as a certainty that no man ever served so long and effectively in a similar capacity in behalf of any breed of livestock as did R. J. Kinzer in his virtually life-long dedication to Herefords. The breed, years later, continued markedly to reflect this contribution.

# F. R. Carpenter

HAYDEN, COLORADO
1886-1980

FARRINGTON REED CARPENTER was his full and formal name. It was known to perhaps but few. To his long-time neighbors in northwestern Colorado, and to his Hereford-industry friends across a broad section of the plains and mountain country, he was "Ferry" Carpenter, a contraction which he liked. It was a name which became associated with the ranch-country town of Hayden, Colorado, in 1907, when he was only 21 years old, and it was his address 73 years later, when he reached the end of the trail at the age of 94, only a month after his induction into the Hereford Heritage Hall's Honor Gallery.

Ferry Carpenter throughout a lifetime of accomplishment was his own man. He never was one to seek the easiest path or to back away from a challenge. His independent spirit led him time after time to go his own way. If he ever regretted this instinct, there is no indication of it.

As the years passed, he seemed to revel in the fact that someone once had referred to him as "the orneriest man in the West." Nor did his advancing years mellow him. "When you get to my age," he once related as his 90th birthday approached, "friends are always saying, 'Ferry, you're looking well.' After about so long you get awfully tired of hearing 'em lie." Then, with eyes a-twinkle, he chuckled. This, too, was characteristic.

He seemed to thrive on controversy, but his successes outweighed his failures. And he never seemed to be concerned that others might disagree with him. Nor worry about it. As when he was hung in effigy by a bunch of angry coal miners, or when he defended an alleged cattle rustler at a time when this was not a very popular thing for a young lawyer to do. Yes, Ferry Carpenter was a law school graduate and a practicing lawyer, but in a small country town like Hayden, he once said, there wasn't much for a lawyer to do.

But perhaps the longest-lasting of any of his disagreements was with those engaged in the registered Hereford business who chose to fit and show the product of their herds at the fairs and shows, and to sell them at the traditional auction sales, usually with much emphasis on pedigree. Located as he was in a region noted for its production of well-grown-out commercial cattle, he chose to produce the kind of Herefords which he figured those commercial ranchmen wanted, without regard to the fancy and folderol which he thought was all too common in certain purebred circles. Never mind, his actions seemed to indicate, that he was considered a crank by some because of his insistence on developing Herefords that he thought were the best regardless of what the other fellow thought.

"Our customers always have demanded big Herefords," Carpenter stated just a year or so before his passing. He went on: "We sell a large number of bulls to commercial cattlemen in Arizona, New Mexico, Texas and southern California. These are what we call warmer-weather states and we feel that cattle produced in warm climates are somewhat smaller than the colder-states' cattle. Our cattle, with their size, are needed to help the warmer-climate breeders keep up the size of their herds' output."

By hewing consistently to that line, Ferry Carpenter avoided the fads and frills of beef cattle conformation and size, never worrying for a minute as to whether fellow breeders or breed association officials approved or disapproved. Ferry Carpenter was first, last and always a rugged individualist. And it could not have been other than highly gratifying to him in his latter years when his philosophy came to be so widely endorsed in the Hereford business, and the beef cattle industry in general. But, even so, his sense of satisfaction was pretty well hidden.

Except for a teen-age mishap, Ferry Carpenter might never have become a Westerner. Born in the Chicago suburb of Evanston, Illinois, he was in the midst of a typical Midwestern boyhood when the incident occurred which changed the course of his life. His father, a New Englander who was in the shoe-manufacturing business, had given him a new pair of ice skates for Christmas. What was more natural for an outdoor-type, 13-year-old than to jump at the chance to play hockey with his neighborhood mates. He was

*Ferry Carpenter first came to national attention in the 1930's when he was appointed by President Franklin Roosevelt as first director of the Grazing Bureau of the Department of the Interior under the Taylor Grazing Act. In that position he became widely known among ranchers and spoke frequently at livestock association conventions, as in this instance.—Saxton photo.*

spiked by the skate of another boy, infection developed, and he was forced to miss several months of school.

When spring came, his mother decided that the fresh air and wide open spaces of the West might be good for him, so she made arrangements for him to spend some time at the ranch of John B. Dawson, a family friend, whose place was near the little town of Dawson, New Mexico. There Ferry rode burros, chased coyotes and did other things that curious boys usually do. But most of all, he learned to love the West. And returned again and again to work at Dawson's ranch for his board and room during his summer vacations.

After selling out to a coal-mining company a little after 1900, Dawson found a new ranching location in northwestern Colorado and bought raw, undeveloped land in the east end of the Hayden Valley, not far from the little town of Hayden. Ferry first worked there during the summer of 1905. He continued to be fascinated by the West, and especially liked this high country with its elevation of over 6,000 feet. He could not imagine, he later said, living anywhere else. But he probably little dreamed, as he rode west in 1905 from Steamboat Springs on top of the stage coach next to the driver, that some day he would own this very ranch.

While Ferry's family was by no means wealthy, his father's income was sufficient to provide comfortable living and good educations for his children. So before the son could become forever rooted in the West, his father insisted that he enroll at Princeton University in New Jersey. He acquiesced, but as he progressed toward graduation he continued to work during the summers at the Dawson ranch. And during this span, too, he homesteaded 160 acres for himself.

He once related how he acquired his homestead in Colorado's Routt County: "It was early summer in 1907 when I put a saddle on a horse, packed some provisions, tied them and a rifle on the saddle and went north from the Dawson ranch to explore the possibility of finding some land to homestead. I was lucky because about 10 miles north of the Dawson ranch, on Elkshead Creek, I discovered the only spring of water in that part of the country." About the same time one of Ferry's good friends, Jack White, filed a homestead claim on a nearby 160 acres, so the two of them had 320 acres, with the only available water for miles.

In 1909, the year Carpenter got his Princeton degree back east, he and Jack White attended the dispersion sale at Parshall, Colorado, of the registered Hereford herd of Elias M. Ammons, later to be the state's governor. Having arranged to borrow some money from Ferry's father, they bought 25 cows for a total of $2,500, thus founding the herd of the Carpenter & White Hereford Company. The cattle were run on the two homesteads and the surrounding free public range. Only two females–in 1915–were added in the more than 70 years of Ferry Carpenter's continuous operation in the development of the long-bodied, fast-growing cattle which represented his ideal. No available record remains of the bloodlines represented by the Ammons cows, but that they probably were better individuals than the average of their times is indicated by the fact that Ammons was an exhibitor at the early Denver stock shows, including the very first one in 1906.

Even though he had homesteaded his first land and started the Hereford herd in partnership with White, Carpenter's father insisted that he proceed next to get a law degree, urging that it would be an extra tool to use in building a business career. So it was that in the fall of the same year the Ammons cows were bought he went to Harvard University's Law School. But as soon as he acquired his degree he headed back to Hayden where he carved a unique place for himself, both in public life and his profession, as well as in the Hereford business.

Realizing that he had much to learn about the Hereford business, he sought the advice of such leading breeders as John E. Painter, an Englishman who was one of Herefordom's pace-setters in the earlier decades of the 20th century, and eventually a two-term president of the American Hereford Association. "Get some of the blood of the bull, Anxiety 4th, that Gudgell & Simpson imported from England," Painter told him. As a result, the firm of Carpenter & White became the owners of Beau Blanchard 64th. This bull, bought from Jesse Engle & Sons of Sheridan, Missouri, in their 1919 auction, was of strong Anxiety 4th ancestry. Described in that Engle sale catalogue as a wonderfully burly bull, his sire was Beau Blanchard, a noted son of Mousels' Beau Mischief, while his dam's close-up ancestors included such Gudgell & Simpson luminaries as Beau President, Beau Brummel and Lamplighter.

Mainly by retaining as herd sires bulls produced in the herd, this brings the Carpenter & White operations down to about 1924 when hard times beset the nation's economy during the post-World War I period. White decided to pull out of the firm to become manager of a ranch in Colorado's North Park owned by Claude Boettcher, a Denver financier. Carpenter continued his own operations thereafter under the name of the Dawson Cattle Company.

John Dawson had sold his holdings in the early 1920's to the Victor-American Fuel Company from which Ferry leased the 2,400-acre property for a period of 19 years. In 1944 he acquired ownership of the ranch through foreclosure on fuel company bonds which he bought up when they had not been redeemed.

Minimizing his practice of law, which was never a dominant element in his activity, Carpenter intensified his efforts in succeeding years in seeking to approach in his Hereford program his own personal ideal in the matter of type. This was considerably different from the smaller and blockier kind increasingly in vogue by the 1930's and afterwards. His goal, as he once pointed out, was to produce more beef per animal unit without appreciable diminution of quality. He sought big bulls which he believed would hasten his progress, being more concerned, he once stated, with their performance than with their pedigrees.

He found a sympathetic listener to his plans in Dr. R. T. Clark who, in the 1930's and later, was a leader at the U.S. Range Livestock Experiment Station at Miles City, Montana, where a program of breeding larger-framed Herefords was under way. Clark was impressed by the Carpenter program which was pointed in the same direction, and the upshot of their meeting was that two Line 1 bulls from the Experiment Station were loaned to the Coloradoan. This infused into the Carpenter program the Advance Domino 13th influence, which had gone to Montana from Fred C. DeBerard's herd at Kremmling, Colorado.

Still another source of the genetic input of larger-size cattle came from the herd of John and Mary Crowe of Millville, California, where performance records had been maintained since 1942. This was through the 2,200-pound Super Prince, a descendant of Super Superior, which traced through Mousel Bros.' Beau Mischief and the G. & S. herd bull, Bright Stanway, to Anxiety 4th. And so it went, with the heavy emphasis always on performance in making selections of herd bulls which matured at a ton or more and herd matrons in pasture condition averaging 1,250 pounds and above.

In more recent times, Carpenter had operated pretty much a closed herd of around 150 cows, with top bulls of his own production in service. One of these was Lad 38, which traced back through several generations of Victor Dominos to Anxiety 4th. This bull was the progenitor, through Viscount 4th of the FRC Boys, Sirs, Kings, Counts, etc. A Carpenter year-old bull double-bred to Viscount 4th topped the 1979 sale at the Midland Beef Cattle Performance Testing Center at Billings, Montana, at $29,000. It had been Carpenter's practice in his later years to send 10 bull calves annually to this test station which made it readily possible for him, as well as his customers, to compare the performance of his herd's output with that of other performance-oriented breeders who were utilizing the station's highly regarded facilities.

Another Carpenter line of recent prominence descended from a son of SH L1 Advancer 934, a bull bred by Scott Holden of Absarokee, Montana, and a descendant of the Line 1 family from Miles City. A year-old Carpenter-Williams bull calf from this sire line topped the 1980 Midland Test sale at $27,000, and of him, Leo McDonnell, manager of the testing station and of the sale, said: "He has the most complete performance of any bull ever tested at the Midland Test Center."

Other calves of these same lines have sold in consignment sales at up to and above $20,000, and a remarkable demand for them developed at the ranch during the latter 1970's. For example, 40 head were sold in the Carpenter-Williams auction at Hayden in October, 1979, for an average of $3,654. Fifty-six head, averaging around 10 months old, were sold in October, 1980, at a rate of $2,376.

Reference to Carpenter-Williams deserves a word of explanation. In 1958, learning that Ferry needed someone to run the ranch, Melvin Williams, who was employed at a neighboring ranch, expressed an interest and was hired. The arrangement was so compatible that in due course the operation became a joint venture. Always sympathetic to the ranch employee's viewpoint that

*Although enfeebled by his years, Ferry Carpenter continued his activity with his Herefords well toward the end of 1980 when he had passed his 94th milestone. Here he is shown casting an eye toward one of his herd bulls.—Carpenter photo.*

job security frequently is shaky, Carpenter for years allowed Mel Williams to run 15 to 25 of his own cows in the herd, and sell his bull calves in the auction. This was an appreciated supplement to the customary monthly earnings. Carpenter also helped provide for Mel's future by selling him some land at a reasonable price and on easy terms.

So it was that for more than a score of years these men worked together, with the same performance goal in view. Carpenter had maintained records through Performance Registry International since the mid-1950's, and eventually also enrolled in the American Hereford Association's Total Performance Records program. He had emphasized birth weights, weaning weights and yearling weights, plus measurements, for years, adding further testing elements as they came into use. He relied on weight per day of age and measurements in selecting both bulls and replacement heifers, using their pedigrees to a lesser extent. And who can deny that he gained the results which he sought?

The sales practices at the Carpenter ranch also differed markedly from those used by most purebred herd owners. Until the mid-1970's he always sold his bull calves at the ranch by segregating them into lots according to their weights per day of age. All which figured three pounds or more were lotted together and priced accordingly. The other lots went from 2.9 pounds down to 2.1 and were priced lower as the recorded weights decreased, with all calves in each lot selling for the same price. There was no competitive bidding. Each person wishing to buy drew a number and made his selection according to the number he had drawn. Duplicate selections were settled by lot. It is said that many a Carpenter customer during this era found this to be a bargain spot for buying bulls, frequently in the sale program's earlier years at prices not exceeding $500.

During the final five years of his life, Ferry Carpenter revised his sale procedures, but they still differed from those of other breeders. Beginning in 1976, auction sales were held at the ranch each October. A base price was put on each bull and there was competitive bidding when more than one prospective buyer wanted the same bull. Rather than a professional auctioneer, Mel Williams acted in that capacity. Sales were held outdoors, with the bed of a pickup truck or a wagon serving as the auction box. There was nothing high-powered about the event; rather, it was a distinctly soft-sell procedure which proved popular with the buyers. But it got the job done as indicated by the averages recorded during this last five-year span.

Despite his well-known aversion to the showring, Ferry Carpenter probably would have found a sense of perverse satisfaction in the fact that an FRC heifer which he bred became senior and grand champion female at the

*Ferry Carpenter was a "working" rancher, and pulled his share of the load throughout his many years in the Hereford field. Here, rope in hand, he was ready to head out toward some action.—Carpenter photo.*

1981 National Western Stock Show in Denver, an event which marked the Centennial of the American Hereford Association with one of the largest and best displays in modern breed history. Too bad, friends thought at Denver, that he couldn't have lived a few weeks longer to have experienced this victory. For it was a fact that he had not changed his ideals to meet showring standards. Rather, the showring standards had moved in the direction he for so long espoused. This was evident in the fact that the big-framed, early spring yearling which scored the victory was the tallest Hereford female in the show.

By no means was ranching and Hereford-breeding, plus the practice of law in Hayden, the only activity of this vivid, earthy, perceptive man. He not only was a talented and acute observer of local, state and national affairs, but during his well over four-score-years-and-ten he had served Hayden, Routt County and the state of Colorado in many roles: town and county attorney, school board member, state representative, Colorado's first state director of revenue, and others.

One of these "others" was his national service as the first director of the grazing bureau under provisions of the Taylor Grazing Act. In this capacity, during the 1930's he brought a large measure of order out of the chaos and conflict which had long prevailed in connection with pasturing livestock on the national forests and other federally-owned lands throughout the West. He was an appointee of Secretary Harold L. Ickes of the Department of the Interior during the first administration of President Franklin D. Roosevelt, and his clashes with Ickes became headline news. Ickes, whose reputation as an "old curmudgeon" was well deserved, met his match in the maverick Carpenter who proved that he could be just as feisty.

On an occasion which he once related, when he and Ickes were meeting with a group of stockmen in Nevada, he commented that "trying to speed up a bureaucrat is like throwing sand into the gear-box of a machine—it stops him cold." The shocked Ickes was so upset that he fired Carpenter on the spot. But the stockmen, who appreciated Ferry's efforts, raised such a protest that

Roosevelt reinstated Carpenter over Ickes' objection. "I got things done, but I got fired just about every time I spoke out," Carpenter once related, adding: "Finally, in 1939, I'd had enough so I came home and concentrated on a different kind of bull–long-bodied Herefords."

What has been written here by no means tells all of the Carpenter story. There are many untouched details such as his acquaintance with President Woodrow Wilson, how he sold bulls in the hard times of the early '30's by hauling samples by truck as far distant as San Francisco and leading them into a prospective buyer's downtown office, and how he helped get the first public school started in Hayden. The latter occurred at a time when there were only two women in the whole territory, both of them married. Ferry and other young men thereabouts thought a school might bring a schoolmarm or two to the little town. It did, and he married one of them, a Kansas University girl from Lyndon, Kansas. One of the sons resulting from that union was Edward Carpenter, of Grand Junction, Colorado, who accompanied his father to Kansas City to witness the latter's induction into the Hereford Honor Gallery. Perhaps all of these things, and more, will be related in more detail in Ferry Carpenter's autobiography which was virtually completed at the time of his passing, and which, it is hoped, will be published in 1982 or shortly thereafter.

There can be no doubt that Ferry Carpenter was a unique "do'er" and probably the most colorful personality on the western slope of Colorado for well over a half-century. Perhaps the scope for that observation might be expanded to cover the entire state–or the entire Rocky Mountain West.

In any event, he was his own man. He proved it again and again. There was a popular song a few years ago entitled "I Did It My Way." As much as any man can, Ferry Carpenter did. And today's Hereford industry reflects his influence and convictions.

*Had he lived another month or so, Ferry Carpenter would have been highly gratified by the grand championship victory scored by a heifer of his production at the 1981 Denver stock show. Named FRC Yampa Donna 42d, she was exhibited by Las Vegas Ranch, Prescott, Arizona. Larry Stark of that firm is at the halter.–AHA photo.*

# Hayes Walker, Sr.

KANSAS CITY, MISSOURI
1876-1944

THE name of Hayes Walker, Sr., is indelibly linked with the history of the Hereford breed in the United States. Born and reared on a farm at Walton, Kansas, he left home at the tender age of 16 to become a printer in the shop of the Kansas City *Daily Drovers Telegram*, the region's livestock industry newspaper. So assiduously did he pursue this task that a few years later he became a fieldman for that publication, and years later recalled that his very first trip in that capacity was to Robert H. Hazlett's Hazford Place at El Dorado, Kansas, where a registered Hereford herd had been established only two years earlier.

After a few years of experience in this capacity, Hayes Walker and two or three other young men pooled their resources and undertook the publication of *The Breeders' Special*, which was somewhat more specifically directed toward the purebred field. But this still was not quite what he wished to do. So he withdrew from that firm to establish *The American Hereford Journal*. On May 1, 1910—in a year when Hereford registrations totaled a mere 23,614 head—the first issue went into the mails—10,325 copies, it is recorded.

Hayes Walker's words written in 1935, carry an undertone of the vision and determination which accompanied his new undertaking. He wrote:

"For several years I had had a conviction that a publication devoted exclusively to the Hereford breed could be made of immense value to the industry it served, and as a consequence a profitable business venture. My first inclination to the Herefords came during my early years on the Kansas City livestock market, and continued while on my job as a livestock paper fieldman. More and more as I came to know about Herefords and Hereford breeders of the Central West, the more firmly convinced I became that the Hereford eventually would become the most popular of all the beef breeds. As my acquaintance with the breeders grew, their spirit of enterprise and aggressiveness appealed to me—and provided the incentive to me to make the venture in such a highly specialized field."

There were at that time, of course, general livestock and state farm publications which carried news and publicity on all kinds and breeds of livestock, but the idea of a one-breed magazine was new—at least as far as this visionary young man knew. Even though it was revolutionary, his conviction grew that such a publication would meet a need and give impetus to the Hereford breed's advancement. The lack of a medium to place the focus on the breed's achievements, and spread the word across the entire country, he considered to be a serious handicap to breed progress, and so the die was cast.

*At the 1919 sale of Wallace & E. G. Good at their Good Donald Farm near Grandview, Missouri, Hayes Walker stood between two Hereford compatriots. The one at left is John M. Hazelton, for many years editor of* The American Hereford Journal *and breed historian, while at the right is W. L. Lacy, a Kansas City Hereford breeder.*—Hereford Journal *photo.*

"We started with a larger equipment of hope and faith than of finances," he once stated. "A second mortgage on my home raised a few hundred dollars," he added, "and this, with a few thousand dollars worth of printing due from the sale of my interest in *The Breeders' Special*, furnished the financial foundation."

Although rough financial weather was encountered not only at the outset, and at several times along the way, never once did Hayes Walker alter his straight-ahead course. Not once in its twice-a-month schedule under Walker ownership did the *Journal* fail to appear, regardless of what personal sacrifice might have been involved. He sincerely disclaimed any personal credit for the breed's advance during succeeding years, asserting that he merely told through the pages of the *Journal* of what the Herefords themselves accomplished. Others, though, had different ideas on that score.

In addition to the material things conveyed through the *Journal*'s pages, he radiated personally and through his writing a brand of optimism and inspiration which encouraged breeders when beset by difficult times to "hold on, hold on." No one can say how many "comebacks" might have been credited to his expressions of faith and his encouragement of hope. One thing above all in his business life: He was for Herefords–first, last and all the time. He was loyal to the breed and breeders alike, found great zest in being among them, and few were the weeks during his 34 years as the *Journal*'s publisher that he did not attend a show, a sale or a meeting somewhere–right up to and including the final month of his life in 1944.

Among the many who testified to Hayes Walker's noteworthy contribution to the advancement of the breed's interests was Fred Reppert, whose rise to eminence among livestock auctioneers almost exactly coincided with the career of the *Hereford Journal*'s founder. In writing some 25 years after the beginning of the role of the *Journal* in promoting Hereford progress, Reppert said, "I doubt if many breeders today appreciate the important part that the *Journal* played in building up this wonderful breed in the United States." He continued:

"To produce good cattle and win high honors for the breed in the showring, in the feedlots and on the markets, and not let the world know about it, would be like hiding one's light under a bushel. For 25 years the *Hereford Journal* not only has sponsored the Hereford breed and told the world of its superior qualities, but twice every month, through its columns, it has gathered the Hereford breeders together for a round-table discussion of means by which the breed might be improved and its progress furthered, and how the pitfalls that have obstructed the advancement of other breeds of livestock might be avoided by producers of Herefords. . . .

*This picture emphasizes how greatly the Hereford breed expanded with the advent of* The American Hereford Journal. *The first 24 issues, comprising its first year of publication, are in the small volume at left. The two volumes at extreme right include the 24 issues published during its 10th year—from May, 1919, through April, 1920. In the* Journal's *first year, 1910-11, a total of 1,203 registered Herefords were sold at auction for an average of $160, whereas 10 years later the average was $491 on 19,095 head.*—Hereford Journal *photo.*

"The reports of shows and sales published in the *Journal*, as well as Hayes Walker's inspirational articles in its columns, have been of untold value to the breed and to the breeders, and have made many converts to the Hereford cause. There has never been a time when Hayes Walker faltered in his support of the breed, or failed in his optimism over the future. Even when that future looked dark and gloomy to the breeders, and they might well have lost heart, it was the *Hereford Journal* that reawakened their courage and renewed their enthusiasm by its confidence, staking as it did, everything it possessed, even against the advice of its friends and bankers, on the future, and carrying on even in the face of heavy financial losses. Yes, the Hereford industry owes a great deal of its success to Hayes Walker and the *Journal*."

Colonel Reppert's testimony given a quarter of a century after the *Journal*'s beginning is evidence that Hayes Walker's goal, as set forth on the first page of the initial issue, was more than achieved. Portions of that statement follow:

"We believe every man who reads *The American Hereford Journal* regularly will be a more successful breeder, a more ardent worker in the interests of the breed, and a stronger advocate of the merits of Hereford cattle as a result. These benefits, we believe, will come not so much from anything we may be able to say in behalf of the breed, but from the information we shall be able to gather from the individual breeder and present to the breeders as a whole through the columns of this publication.

"*The American Hereford Journal* will be a medium through which the breeders of Hereford cattle may exchange ideas and experiences. By means of it men just starting in the business may have the benefit of the expensive lessons which older breeders have learned by experience. Through its columns problems which confront one breeder will be solved by others who have worked out the solutions by years of study and effort. In short, *The American Hereford Journal* will be a clearing house for the Hereford Breeders of America. . . .

"We venture to say that it will not be possible for any Hereford breeder to collect from all other sources combined the volume of valuable Hereford history, information, literature and live news that will be found in the columns of this publication. It will be a mine of information, a library of reference, a source of inspiration for every Hereford breeder." Thus Hayes Walker wrote.

It may be of incidental interest that this first issue contained a total of 24 pages.

Commendations from Hereford breeders across the country flowed in during the ensuing weeks, several of which were published in subsequent issues of the *Journal.*

John E. Painter of Roggen, Colorado, said: "I think your paper will be a medium of bringing breeders closer together to work for the betterment of the breed and the advancement and extension of it into new territory."

Robert H. Hazlett wrote from El Dorado, Kansas: "The first issue is splendid, and your salutatory is about the best thing of the kind I have ever read."

Col. R. E. Edmonson, auctioneer and Hereford rancher at Claude, Texas, stated: "If you never fall below this first issue, you are a winner."

C. R. Thomas, then secretary of the American Hereford Cattle Breeders' Association wrote: "I wish to congratulate you. . . . If your future issues are as newsy as is your first, I am satisfied that success awaits you and that no breeder can afford to be without this paper."

Warren T. McCray, owner of Orchard Lake Stock Farm, Kentland, Indiana, and a director of the Association, wrote: "I want to congratulate you. . . . I am sure of the need for just such a publication and I have faith that its influence will be a strong factor in the advancement of the popularity of the breed."

Another Association director, Oscar L. Miles of Fort Smith, Arkansas, owner of Point Comfort Place, wrote: "I have examined the first issue rather critically, and approve it entirely. I believe it will be a great enterprise."

The then-current president of the Hereford Association, C. N. Comstock of Albany, Missouri, stated: "I think your magazine is a dandy, and am

*This 1920's picture of Hayes Walker was embellished by an artist's sketch to depict his interest and activity in connection with Herefords. His right hand extended the cover page of an issue of the* Journal *which featured a picture of Richard Fairfax, a bull sold in 1919 by L. A. Pinard of Wessington Springs, South Dakota, to Ferguson Bros. of Canby, Minnesota, reportedly for $50,000, a record which stood until post-World War II times. The bull shown above is unidentified.—Lacy sketch.*

delighted that we have a journal published in the interest of the Hereford breed."

After reading the first issue, James T. Boney wrote from Cairo, Missouri: "If this is a sample of the issues to follow, the *Journal* will be of inestimable value to Hereford breeders."

"No Hereford breeder can afford to be without it," wrote Mousel Bros. of Cambridge, Nebraska, while John Letham, then of Lake Geneva, Wisconsin, expressed his hearty approval on every point except the 50-cents-per-year subscription price. "It is entirely too low; any man with Whitefaces would cheerfully give double."

C. M. Largent wrote from Merkel, Texas: "I feel that it will inspire the old breeders and add many new ones," and from Adrian H. Hewes of Sundance, Wyoming, came the comment that he had "wished for a publication of this nature for some time." Samuel Drybread wrote from Elk City, Kansas, that he saw "no reason why the *Hereford Journal* should not grow and wax strong."

Many other similar comments appeared in subsequent issues, and the success of the venture was assured.

Hayes Walker in 1914 planned and executed the first summer special edition of the *Journal*, the forerunner of the Herd Bull Editions which reached the 58th in this impressive series with the 1981 issue. This was an innovation when the series was initiated, but in the decades which followed it was the rare publication in either the cattle or swine field which did not publish a similar summer edition. During Hayes Walker's lifetime, this special grew in size to an issue which contained 484 pages, and later, during the post-World War II times, advanced to well over twice that size.

Following the founder's death, his son, Hayes Walker, Jr., carried on successfully as publisher of the *Journal* until its sale in 1961 to the American Hereford Association, of which it then became the official publication.

For a great many years, Hayes Walker encouraged Hereford breeders onward and upward through his personal column in almost every issue, published under the general heading, "In My Opinion." The tenor of these messages is indicated by individual topics, such as the following: The Unconquerable Hereford, A Source of Pleasure and Profit, Most Beautiful of All Utility Animals, There Is No Stopping Place, Quality Rather Than Quantity, In Debt to the Hereford, Enthusiasm Necessary to Hereford Success, Loyalty Is a Hereford Characteristic, The Value of a Definite Plan, No Success From Half-Hearted Efforts, The Guarantee of Performance, The Mutuality of Hereford Interests, and on and on and on.

Hayes Walker continued active until the very last. In the issue immediately before his passing he wrote a front-of-the-book feature, in the encouraging and inspirational vein which long since had become his hallmark, entitled, "Herefords Must Lead Industry or Fail."

There is no doubt that his name belongs in the list of those who wrought faithfully and earnestly for the advancement of Herefords and Hereford interests throughout America.

# E. H. TAYLOR, JR.

VERSAILLES, KENTUCKY
1829-1923

FEW men make their entry into the Hereford field, or begin any other business endeavor, at the age of 84 years. Fewer still, regardless of the field of activity in which they may be involved, record any significant accomplishment after that milestone has been passed.

One who beat the odds was Col. E. H. Taylor, Jr., of Frankfort and Versailles, Kentucky, who, at that age, bought his first registered Herefords in 1913. Then, in 1914, from his Kentucky neighbor, Col. W. H. Curtice of Eminence, he bought a substantial foundation herd of 20 head. During the ensuing nine years, until his death at the age of 93, he cut a swath in the Hereford world rarely matched, regardless of the time span involved. It is true that Colonel Taylor possessed enormous financial resources which he poured liberally into this enterprise, but even more important was the dedication to excellence which was manifested again and again in the development of the herd at his place at Versailles, which was known simply as Hereford Farms.

Possibly the vitality and energy that characterized his whole life played some role in his Hereford decisions. He was a prominent figure in Kentucky banking circles before the Civil War. He was for many years mayor of his home town of Frankfort, and served as a member of the state legislature. His first and greatest business success was in building a distillery empire, making such famous brands as Old Hermitage, Old Crow, Pepper and Old Taylor whiskey. The same imagination and drive which resulted in those successes again were brought into sharp focus during his later years in his Hereford enterprise.

The key to the immediate success which attended the Taylor effort was his purchase in the spring of 1914 of the two-year-old Beau Perfection 24th from Colonel Curtice at $12,400, the highest price on record to that date paid for a Hereford in the United States. The same deal included 19 females, all of Beau Donald descent, and comprising all of the cows in the Curtice herd carrying the service of this young bull, some of them members of the 1913 Curtice show string. Beau Perfection 24th was sired by Perfection, grand champion bull at Kansas City in 1900, he a son of the 1899 Kansas City event's grand champion Dale, and tracing to Earl & Stuart's Garfield, while his dam was a double granddaughter of old Beau Donald. It was the Perfection-Beau Donald nick which brought fame to the Curtice herd. Shown as a calf and yearling by Curtice, Beau Perfection 24th stood second in class to the Overton Harris bull, Repeater 7th, at the American Royal show in both 1912 and 1913, and was ranked as a "comer" by almost all who saw him.

In fact, this prospective herd-header was pronounced by numerous contemporary Hereford students and judges as possessing the greatest combination of show and breeding qualities to be found in any young Hereford bull of his time. It was freely predicted that should his value as a show bull be in any way diminished, he would still retain his sterling promise

*Colonel Taylor's Hereford establishment almost immediately became famous as one of the industry's most impressive and imposing. Some of the brood matrons pose here in front of a huge barn.–Hildebrand photo.*

to become a truly great sire. At the time of his purchase by Colonel Taylor, at the age of two years and eight months, he weighed 2,200 pounds.

This description of him was written at the time: "He is wonderfully smooth. He has a head and neck that an artist could scarcely improve upon. Character stands out prominently in every feature. He presents a very pleasing appearance to the eye, with a strong back and loin, and a rump as smooth as could be produced by the chisel of a sculptor. Strength of constitution is shown by his extra large girth and width between his forelegs, and he is as active on foot as a coach horse."

Immediately after his purchase by Colonel Taylor, this great young bull was re-named Woodford, in honor of the Kentucky county in which Hereford Farms was located. And the bull proceeded to imprint his name firmly upon the Hereford consciousness all across America, despite the relative brevity of his career.

But first came a disappointment. The new owner decided to exhibit Woodford to the Chicago International show in the fall of 1914. When an outbreak of foot-and-mouth disease caused the show to be cancelled, Woodford's chance at the grand championship, which the owner sought, was gone. In the opinion of John Letham, a Scot who was considered by old-timers to be a truly great judge: "His equal had never been seen at the International in the two-year-old class to which he was then eligible." And

*Woodford as pictured in a snowy setting at Col. E. H. Taylor's Hereford Farms. Purchased at a record price as his owner began his activity with Herefords, Woodford gained fame as one of the most significant breed-builders of his era.—Hildebrand photo.*

then Letham said prophetically: "If he lives he will create a family of Herefords that will go down in history."

Colonel Taylor swallowed his disappointment that Woodford's opportunity for a Chicago victory was blocked by fate and found zest in the subsequent showyard performance of his herd's representatives, many of which were sons and daughters of this herd sire. For example, in seven of America's leading Hereford shows in 1917, first-prize awards scored by Taylor entries totaled 106. Throughout the remainder of the 'teens, the record was about the same, including also many championship purples.

The get of Woodford also reflected the produce of one of the greatest collections of Hereford females ever assembled in a breeding herd. These comprised the cream of numerous top American herds, including 30 from the famed Beau Brummel-bred herd of W. A. Dallmeyer of Jefferson City, Missouri, plus a shipment in 1915 of some 40 head of the best then available in England. Among the latter was the English Royal champion Clive Iris 3d which became the American Royal grand champion female in 1917, and then went on to become the top-seller in Colonel Taylor's first auction sale in June, 1918, when she fell to the $13,850 bid of W. R. & W. A. Pickering of Kansas City and Belton, Missouri. She thus established a new American record price for a Hereford female. This entire sale of 62 head averaged $3,010 in setting a new world's record for the breed to that date.

By this time Woodford had established himself beyond doubt as one of the great herd bulls of the times, if not of all time. He proved equally impressive as a sire of bulls and heifers. There was a uniformity in his get seldom seen in the get of any bull. Beginning in 1916, the winnings of his sons and daughters, and grandsons and granddaughters were little less than sensational. Blue and purple ribbons were accumulated by the Woodfords to a number that the founder of the Woodford line could not possibly have anticipated.

Best of all was the fact that the breeding performance of his early sons began by now to demonstrate the significant family characteristic of prepotency. After winning as members of the show string, many of them went into herd service either at home or in other important herds. For example, Woodford 6th, one of his first sons, was junior champion at the Chicago International in 1916, then topped the International Hereford sale at $15,100 in going to the Hartland Farms of Sen. J. N. Camden, also at Versailles, for whom he sired a procession of prize-winners and herd-improvers, including grand champions of both sexes in top national competition.

At least a dozen other Woodford sons and grandsons, including the 4th, the 8th, the 9th, the 28th, the 61st, the 63d, the 65th, the 71st, the 125th, the 167th and Woodford Prince, among others, became noteworthy contributors to breed progress, while his daughters and their descendants were prized as brood matrons in numerous herds.

Just when the future loomed most brightly, disaster struck. As a staff member of *The Breeders' Gazette* wrote somewhat effusively concerning Woodford in the style of the times: "He perished in majestic isolation on October 3, 1918, in the dead of the autumn night, enveloped in the giant flames which burned to the ground in a few hours the magnificent 208x68-foot barn which Colonel Taylor had recently erected as a fitting home for him and his remarkable calves at Hereford Farms." Fortunately the calves which shared this facility had been turned out for the night, and thus were spared to carry forward their sire's influence.

Forty years later, Woodford still ranked 18th among the 128 bulls which by then had gained recognition on the American Hereford Association's original Register of Merit. This is doubly significant in view of the fact that the points accumulated to his credit were gleaned during the period when only two shows annually were designated as Register of Merit events–the American Royal and International. It is rather amazing, as well as a tribute to Colonel Taylor's foresight and determination, that Woodford for so long continued to rank so high in this select circle. Three of Woodford's sons also gained Register of Merit recognition, emphasizing again this family founder's remarkable prepotency.

But Colonel Taylor was undeterred by the loss of Woodford. Later in 1918 when in his 90th year, and already possessing Hereford laurels beyond the imagination of most men, he continued to look to the future. Early in the following year, from Herefordshire to Hereford Farms came the highest-priced bull imported to that time from England into the United States. This was His Majesty, bred in the famous Tarrington herd of William Griffiths in Herefordshire, and considered by many of England's leading breeders to be one of the best individuals ever produced in that country. Likewise, his first calves on the other side were said to have been of high promise.

But time, despite Colonel Taylor's seeming immunity, ultimately took its toll three years later and the herd shortly was sold by his estate. A generation and more after his death there were still those who vowed that had Colonel Taylor and Woodford been spared for a few more years, and had a policy of consistently concentrating Woodford's blood been pursued, this bull might well have matched Prince Domino as a benefactor of the Hereford breed.

Indeed, in view of what was accomplished in less than a single decade–about a minute in the march of time–it is small wonder that men have conjectured as to what the Woodfords might have accomplished had the bull lived a normal span of years.

# FRED REPPERT

DECATUR, INDIANA
1877-1946

WHEN a Hereford breeder gets his herd fully developed and its output is ready for sale, his task is only half done. He then has to successfully sell his herd's produce. And that's where an auctioneer comes into the picture. Specifically, that's where Fred Reppert entered the Hereford scene around 1910. During the ensuing quarter of a century, he became by far America's leading auctioneer of Herefords. There had been good auctioneers of Herefords before that time, including such professionals as Col. R. E. Edmonson of Claude, Texas, who himself was a Hereford breeder, but there was no one previously who had dedicated himself and his career so wholeheartedly to the task of purebred livestock salesmanship, and particularly to the merchandising of registered Hereford cattle.

Fred Reppert, the son of an Indiana farmer and farm sale auctioneer, was of Swiss-German descent, an inheritance which occasionally came to light when a bidder of seemingly similar ancestry found himself on the receiving end of a snappy comment delivered in the Colonel's approximation of the Germanic "lingo" with which he had grown up. This usually elicited general laughter and relaxed the assembly. Fred Reppert liked a happy crowd, and did his best always to keep it that way.

The young man from Indiana, by then 31 years of age and well established in his community and state in the auctioneering field, sold his first registered Herefords at the 1908 International sale in Chicago, where he assisted Colonel Edmonson. He continued to work with Edmonson and also with other prominent auctioneers of the times, such as Col. F. M. Woods of Lincoln, Nebraska, while he honed his native ability. He then began to book a few sales for himself, as Edmonson's advancing years made it necessary that he curtail his activities. After Edmonson's death in January, 1912, at a time, incidentally, when he was serving as president of the American Hereford Cattle Breeders' Association, Reppert headed the sale force at practically all of the country's leading Hereford sales until the middle 1930's.

Years later he recalled that one of the very first Hereford sales at which he was the auctioneer was for Cyrus A. Tow, then of Norway, Iowa, in the fall of 1911. Looking back now in the files of *The American Hereford Journal* to the

*Predominant Hereford auctioneer during the decades between 1910 and the mid-1930's was Col. Fred Reppert of Decatur, Indiana. In his heyday he was busy almost every day during the main selling seasons in conducting auctions from coast to coast. This was a characteristic pose as he exerted persuasion and perhaps a bit of cajolery on a hesitant bidder.—* Hereford Journal *photo.*

report of that event, it is found that an average of $181 was recorded. "Not in a good many years," the report stated, "has a sale been held in Iowa that in any way approached this one." Such results and personal acceptance made him the logical successor to Colonel Edmonson, and he carried on with rare enthusiasm and vigor as he conducted record Hereford sales from coast to coast.

His own words, written in 1935, indicated the scope of his activity and the extent of his successes in his chosen field:

"Twenty-five years of hustling, with plenty of thrills, a lot of hard work and a great deal of satisfaction–conducting public sales in every state of the Union, in every province in Canada where livestock production is a significant activity, and in Old Mexico–presiding as salesman at practically every new world's record sale of Herefords–noting from year to year the remarkable progress of the Hereford breed–these are a few of the thoughts which come to my mind in reviewing the past quarter of a century."

Within five years of his decision to make a specialty of selling Herefords, he was the auctioneer when a bull in the 1916 Denver sale brought $5,000 to set a new record for that event. Six weeks later he was the auctioneer as O. Harris & Sons of Harris, Missouri, set a new record sale average for the breed of $1,246, with a top figure of $8,100. He was on the block a few weeks later at the Orchard Lake Stock Farm of W. T. McCray at Kentland, Indiana, when the sale average record was advanced to $1,287, and a top price of $10,000 was established.

From that point onward for several years Colonel Reppert presided as new records were set almost routinely. A bull went through the ring at the 1917 Chicago International at $31,000. E. H. Taylor, Jr., of Versailles, Kentucky, with Colonel Reppert on the block, averaged $3,010, a new high for the breed, in 1918. Mousel Bros. of Cambridge, Nebraska, with the Indiana auctioneer in charge, established a new high for all beef breeds at $4,020 in 1920. In that same year, he sold the top 50 in a McCray sale at an average of $5,575. And so it went as Fred Reppert scaled the heights in his profession.

"Perhaps the greatest thrill I have ever gotten out of presiding at a public sale was at the Gudgell & Simpson dispersion at their Independence, Missouri, farm in the summer of 1916," Fred Reppert stated years afterward. He added: "I was then still comparatively young in the business and it was a high honor to be entrusted with the sale of this famous herd." The animals which he sold there included Domino, Bright Stanway and Lady Stanway 9th, dam of Prince Domino.

On through the 1920's, a period when the post-World War I depression caused prices of all products and commodities to plummet, Colonel Reppert continued with his contagious brand of enthusiasm. He gave his good-humored best, whether the sale was averaging $300 or $1,300. And regardless of the circumstances, he took special pride in the fact that when he "booked" a sale, he would be there to conduct it, "come hell or high water," as he once said. Sometimes this involved travel by hazardous means, or at substantial cost, but at the close of his career he could not remember that he had ever missed a sale with a single exception when illness prevented his presence.

In times when traveling by airplane was a novelty, he chartered a light plane and hired a pilot to take him from Manilla, Iowa, to Atlantic, Iowa, in order to make a sale for A. W. Anstey of nearby Cumberland, Iowa. An old picture shows Reppert and the pilot standing beside an open cockpit Curtiss biplane, each with leather helmet and flying goggles. In later years he traveled extensively by air, and once recalled that he finished a sale in Los Angeles one evening and opened another at St. Louis the following morning at 10 o'clock, an impossibility except for the airplane.

In earlier times he did not hesitate to charter a train if necessary to keep a date. In March, 1912, before the days of airplane travel, he conducted a sale in Kansas, and was scheduled the following day for the sale of W. J. Davis & Co., at Jackson, Mississippi. When he missed his train connection at Springfield, Missouri, he hired a special train, consisting of a locomotive and one sleeping car, to go from Springfield to Memphis, Tennessee, where he was in time to make his connection to Jackson. "This cost me $423.40," he remembered years afterward. After that initial experience, he engaged special trains on several other occasions.

Another sale which involved a major problem in getting there was a mid-winter auction for L. A. Pinard at Wessington Springs, South Dakota. The train got stuck in heavy snow drifts before quite reaching the town, and Pinard sent out sleds to get Reppert and the few Hereford breeders who were on the train. The roads were so badly drifted that even the sleds could not make it. Finally, Reppert and the others had to walk the last three miles to reach the farm. He remembered years later that there were only about 15 breeders at the ringside that day, but added: "We held the sale just the same, and it wasn't a bad sale at that."

Naturally an auctioneer of such dedication and purpose prospered. After a swine auction which he was handling, he was approached to sell a single boar in a forthcoming, widely advertised sale. Reppert suggested that the owner could not afford his minimum fee of $500 to sell a single animal, but the prospective seller countered with an offer of $1,000 to Reppert if he would do the job. The boar brought $21,000.

Then, there was a Hereford breeder who wondered if he could afford

*In the days when an airplane was very much a rarity, especially in rural areas, Colonel Reppert maintained his record of not missing essential connections by chartering this plane in 1919 to get to Atlantic, Iowa, in time to conduct a sale for A. W. Anstey of Cumberland, Iowa. Note the Colonel's leather helmet and goggles, which were necessities for fliers in those "open cockpit" days.—Lewis photo.*

Reppert's price, since he expected his sale to average only around $300. Reppert said, "Will you give me half of what the sale offering will bring above a $400 average?" The breeder agreed. Reppert's share became $22,000, but out of consideration for the breeder who had been so pessimistic, a downward adjustment was suggested, and made, by the auctioneer.

None who knew him ever doubted that Fred Reppert in his heyday was the most colorful livestock auctioneer of his time, and possibly of all time. He played an immensely important role in both the growth of the breed and its improvement. He was a crusader who preached improvement in quality, and his sincerity was so deeply rooted that he came to be considered the greatest salesman the breed ever had. He was a great friend and admirer of the evangelist, Billy Sunday, and they had many traits in common. Each preached his own individual gospel with utmost effectiveness.

This man, who could inspire smiles and laughter in a room filled with sober and glum-faced men through the magic of his words and contagious wit, used the same talent again and again to transform the mood of the crowd wherever he might be, but most of all the crowds at Hereford salerings. Forrest Bassford, a field representative for *The American Hereford Journal* for several years, once wrote: "In my opinion, he was the greatest natural-born salesman in the livestock industry of our time. His life, his joys, his sorrows, his whole being and existence were selling."

Bassford went on: "His penetrating, powerful voice and piercing, attention-getting whistle were familiar all through the United States and Canada. His practical selling psychology exerted a greater influence, in my opinion, in

*Fred Reppert's good humor and contagious smile were widely recognized trademarks during the era when he was the leading auctioneer of Whitefaces.*

spreading the gospel of improved livestock than was exerted by any other man of our time—perhaps of all time."

He possessed what could be described as a "booming" voice. He was in his prime a quarter-century before public-address systems were adopted, but he never needed any mechanical or electronic aid in making himself heard, no matter how large the crowd.

He, too, was a man who enjoyed a good joke. On a warm day he was conducting a sale for J. J. Cahill & Sons at Fairfax, Iowa. Cyrus Tow, the man for whom Reppert conducted one of his very first sales, fell asleep in his seat at ringside. After a number of bids had been made on a cow in the ring, Reppert said in his stentorian tone, "Sold, to Cy Tow," who then awoke with a start. But he was a good sport and took the cow. Later on, Mr. Tow had the last laugh when she dropped a calf that became a grand champion steer.

One other point that deserves mention concerns the personal interest he took in promising young men entering the livestock newspaper field. His concern for their progress was unparalleled in his era. He was an advisor to many newcomers and willingly lent a helping hand when it was most needed. A personal experience may be pardoned. In an emergency situation, this writer, then only a little past his 21st birthday and almost entirely inexperienced, was sent out from the *Hereford Journal* office to fill in for the regular fieldman in "working the ring" at a Nebraska sale. The novice did the best he could, but with considerable misgiving about the result of his effort.

A few days later, the young man's employer, Hayes Walker, Sr., publisher of *The American Hereford Journal*, received a letter from Fred Reppert which all too generously commended the ring work of the newcomer. And the young man, now grown old, never forgot that "pat on the back," however undeserved it probably was. Indeed, he still has the letter which Hayes Walker passed on to him. But that's how Fred Reppert was. And why he was still remembered long afterward by those who had the good fortune to cross paths with him.

# LESTER D. WIESE

MANNING, IOWA
1898-

SOME men are recognized by their peers for dependability as solid as the proverbial rock of Gibraltar. Lester D. Wiese is one of them. Similarly regarded in the Hereford field for a great many years has been the product of a family herd in west-central Iowa which, as of the beginning of the 1980's, is responding to the ministrations of the fourth successive generation of Wieses. Little wonder that Lester Wiese could look back with considerable satisfaction as he was inducted in November, 1980, into the Hereford Heritage Hall's Honor Gallery.

The Wiese family's Hereford activity in the Hawkeye state began with Edward Wiese, who was born at Davenport, Iowa, and who when only a year old traveled in 1870 by ox-drawn covered wagon with his father, D. F. Wiese, and family to the little settlement of Westside, a few miles west of Carroll, Iowa. There D. F. Wiese bought 80 acres of farmland on time. His grandson, Lester, related many years afterward that his grandfather walked 20 miles to borrow money at an interest rate of 24 percent per year to put in his first crop of wheat on that land.

Diligence enabled the family to prosper in its new location and by the time D. F. Wiese died he had accumulated around 1,000 acres, plus other property. Ed Wiese, by then 18 years old, assumed full responsibility and guided the

*Wiese entries which, by 1981, had competed for more years consecutively than the entries of any other beef-cattle exhibitor at the National Western Stock Show, exhibited this string to first-place in the class for "best 10 head" at Denver in 1969. Lester Wiese, senior member of the firm, stood at the left, while a son, Gene Wiese, was at the right.*

activities of the rest of the family. As a youngster he had herded cattle on the unfenced prairie for his father and seemed especially to like the livestock end of the farming business. So when he took over the handling of his father's estate he raised cattle and fed steers, in addition to raising the feed crops. Ed Wiese branched out on his own four or five years later, but continued to live on the home place to run it for his mother until the estate was divided shortly thereafter. It was during this period that Lester D. Wiese was born and introduced to the livestock-farming way of life from which he never strayed.

The indoctrination of Les Wiese in the direction of the Hereford breed began early as his father, Ed, recounted his own boyhood experiences and observations on how well the Whitefaces responded to feed and forage in comparison with cattle of other colors as he herded them in earlier times at Westside. They were, his father told Les, both hardy and docile, and they gained weight more readily than any other kind.

So strong had been Ed Wiese's belief in the white-faced breed that in 1894 he drove a team and buggy from Westside to Guthrie Center, Iowa, a round trip of around 100 miles, to buy his first Hereford bull for use on his bunch of crossbred cows. When Les Wiese was born in 1898, the Hereford color combination already was well established on the family farm, and it is fair to say that he grew up with this breed from his very first year.

With the family matters all settled, Ed Wiese in 1904 moved his own family straight south for a dozen miles or so to the vicinity of Manning, Iowa. There Les grew to manhood and in due course established a nationally recognized name for himself as a breeder of registered Herefords. So well did he like this location that he was still living there 77 years later. And had prospered to the

extent that Ed Wiese's original 240 acres had now grown in extent to around 2,000.

Ed Wiese took with him when he moved his young family to Manning a good commercial herd of Herefords. In addition to continuing to improve this unit through the use of purebred bulls, he fed steers for market on a rather extensive basis. And he continued to be impressed by the performance of both the commercial breeding herd and the efficient weight gains made by the Hereford steers in the feedlot. Thus came a major decision which time long ago proved to be a good one.

In 1912, Ed Wiese, along with his younger brothers, Adam and Elbert, formed the partnership of Wiese Bros., and bought the seedstock for a registered Hereford herd. It is of interest, as with many of their predecessors and contemporaries in the registered field, that their choice of Herefords in entering the purebred business was based upon their successful experience with the Whiteface influence in producing commercial cattle and feeding and finishing steers for the slaughter market. So it was that the Wieses gradually turned their breeding operations from a grade herd to purebreds.

By 1916 they had progressed to the extent that their herd was represented by an advertisement in the summer special edition of *The American Hereford Journal*, a predecessor of the Herd Bull editions of later years. On February 13 of the following year, the Wiese brothers were ready for their first public auction, which was held in a sale pavilion in Manning. It averaged $347 on 57 head. One of its features was their herd bull, Romeo, a grandson of the Gudgell & Simpson-bred Beaumont by Beau Brummel, and out of a dam that was a granddaughter of another noted Gudgell & Simpson herd bull, Dandy Rex by Lamplighter. He was described in sale advertising as being a "thick-meated, deep-bodied bull with a magnificent hindquarter."

Obviously his influence continued in the breeding herd for a time as his successor, Beau Blanchard 29th, which already had been acquired, took over as the new Wiese herd header. This 1915 bull, bred in the already prominent herd of Jesse Engle & Sons of Sheridan, Missouri, from which he was acquired, was sired by the Gudgell & Simpson-bred Beau Blanchard, one of the heralded herd sires of the times, he by Beau Mischief, which gained fame for Mousel Bros. of Cambridge, Nebraska. Dam of this new Wiese bull was a daughter of the equally noted Domino. The intensity of his Gudgell & Simpson inheritance is indicated by the fact that every ancestor in his four-generation pedigree was bred in that famous Missouri herd.

The kind of bull he was from the standpoint of individuality is indicated in the comment about him by John Letham, the Scot who was the highly regarded pioneer field representative of the *Journal*, as follows: "When he meets you with his pronounced Hereford character in a model bull's head, he fascinates you. When he leaves you with his massive quarters and deep twist, one has to admire him."

When time for the next Wiese auction came on May 1, 1918, every bred cow and heifer in the sale, with a single exception, carried the service of this splendid young sire. His influence was clearly evident in the fact that these 42 females averaged barely under $600, with top prices of $1,400 and $1,300, both paid for cows bred to him.

The Wiese course in the direction of Gudgell & Simpson breeding and to that firm's great breed-building sire, Anxiety 4th, now was firmly set. This was apparent in their advertisement in the 1918 Expansion Edition of *The American Hereford Journal* which featured a full-page picture of Beau

*The three men who came to comprise the firm of Wiese & Sons paused in making their rounds at the farm long enough for the photographer to take this picture. Lester Wiese stood between son Gene, at left, and son Sam, on the right.*

Blanchard 29th and pointed out that he headed an equally select herd of cows representing the get of Beau President, Beau Picture, Beau Dandy, Beau Mischief, Beau Blanchard, Domino and Bright Stanway. Without exception these bulls were Gudgell & Simpson-bred history-makers.

Not content with just one good herd bull, Ed Wiese had gone to Mousel Bros.' sale in December, 1917, where he paid $1,650 for Rex Mischief, described by the sellers as "a big, thick sort of bull," which was double-bred to Beau Mischief. This bull, which was used extensively, was not, however, of straight Gudgell & Simpson breeding, which never was a fetish of the Wieses, nor was Lord Dudley, perhaps the first Wiese herd header, calved in 1912, although he was sired by a descendant of Andrew by Don Carlos and out of a dam that was a granddaughter of Beaumont by Beau Brummel, all G. & S. notables.

By that time the foundation program of the herd with which Les Wiese was to be associated for the remainder of his life was well established, and when his father's brothers left the original firm he was ready to step in as his father's partner in the new firm of Ed Wiese & Son. This was in the late 'teens, when the junior member was around 20 years old.

While Rex Mischief still continued as the senior herd bull, Anxiety Mixture 3d was bought in 1924 from Mousel Bros. as a junior sire. Bright Stanway, the last chief herd bull for Gudgell & Simpson and the sale-topper when their herd was dispersed in 1916, was the grandsire of both the sire and dam of this addition to the Wiese battery, and both his ancestry and his individual merit began to be extolled in Wiese advertising a year or two later. Particularly cited were the results of his matings with the Rex Mischief daughters. He left in the herd, Lester Wiese pointed out years later, "a bunch of good cows which nicked exceedingly well" with their next herd bull.

This one was Anxiety Domino, a grandson of the history-making Prince Domino Mischief of Mousel fame, which Lester Wiese selected as a year-old calf in the fall of 1928, making the purchase from Mort Hayes & Son, Lenox, Iowa. The price was $515. The sagacity of the firm's junior member was confirmed when this youngster became a great contributor to the herd's progress, leaving in particular a set of brood matrons which sustained his influence for generations.

Now the stage was set for the next step, and an important one it was. This came in 1933, at the American Royal in Kansas City. Les Wiese was on a search for a successor to Anxiety Domino. Several bulls appealed to him and he bid up to more than $1,000 on the eventual sale-topper before giving up the chase. But good fortune prevailed. Another bull which he liked was Don Axtell 17th, a son of Young Axtell, he a straightbred grandson of Prince

Domino, in the consignment of his breeder, J. C. Andras of Manchester, Illinois. That the junior Wiese was more perceptive than most of those at the ringside is shown by the fact that he paid only a modest $355 for the yearling which was renamed by his buyers and gained nation-wide fame as Intense Domino. It is scarcely too extravagant to state that it was Intense Domino which put the Wiese herd firmly and forever "on the map." His record as a sire of good, big cattle of both sexes which went out to constructive breeders across the country reflected great glory on him and upon the man who saw something in him which other breeders did not see on that sale day in Kansas City. And, most of all, he left a great set of brood matrons in the Wiese herd, as well as two sons which carried on for him in the home herd into the '40's.

Always looking ahead, but in a quietly conservative way, Les attended the National Western Stock Show at Denver in January, 1940, and although not in any immediate need for a bull, ultimately saw a youngster that caught his eye. This was the eight-months-old Battle Spartan 8th, which had ranked fourth in a summer calf class of 39 head for his breeder, Clyde Buffington, then of Crawford, Nebraska. Les watched for him in the Denver sale, to which he had been consigned, somewhat fearful that the price might go too high. But again, good fortune prevailed for the Iowan and the bull calf was acquired at $550, which was just about one-half of the average paid in the sale for the event's top 50 bulls.

Battle Spartan 8th, through his sire, the beef-making Battle Mischief 7th, introduced a new infusion of Beau Mischief breeding into the Wiese herd, while through his dam came additional Beau Mischief influence, along with that of Prince Domino. Since the Wiese breeding herd by then was predominantly of the Intense Domino family, the new bull in due course found many of that bull's daughters in his harem. One of them, in one of Battle Spartan 8th's earlier calf crops, produced the youngster which became the famous Battle Intense. It was precisely to breed such a bull, when mated with cows carrying the previous Wiese bloodlines, that Battle Spartan 8th was bought. This result, however, was beyond all reasonable expectations.

True to the Wiese pattern, his dam represented successive generations of Wiese herd bulls. She, of course, was by Intense Domino; her dam was by Anxiety Domino; the next dam was a daughter of Anxiety Mixture 3d, while next in line was of the Rex Mischief line. Quite a tribute to a progressive breeding program!

Not only was Battle Intense richly bred but he was a bull of superlative excellence, and a contributor to the scale and substance which were prominent features of the Wiese output. The intermingling and intensification of his blood, through both the sire and dam lines, resulted in the development by the '50's of a Hereford family which enjoyed wide popularity. In the following decade the Battle Intenses emerged as one of the breed's top hierarchies. Sons, grandsons and great-grandsons mainly comprised the sire battery, and for years contributed much to the reputation of Wiese Herefords.

In 1949, with Lester's father approaching 80 years of age, his share in the herd was sold to his grandsons, and the firm name ceased to be Ed Wiese & Son and became Wiese & Sons. The "sons" who thus officially joined Lester Wiese in the family operation were Sam and Gene, both widely known in the industry, and the latter destined to become president of the American Hereford Association in 1969. Younger generations of Wieses are now coming along in their families.

Wiese Herefords were seldom exhibited widely on the show circuit but those which did compete acquitted themselves with much credit. The importance and prestige of the Denver show, however, was a constant lure, with the result that, as of 1981, the Wiese herd had been represented continuously at the National Western for a longer period than the herd of any other beef cattle exhibitor there, of any breed, according to Lester Wiese.

When Wiese representatives did enter showring competition, it became the general rule that their presence was strongly felt. Without going into a tedious recital over a period of years, an example of their success may be indicated. One year in the '60's at the Iowa State Fair, in competition with representatives of several other prominent herds, Wiese entries accounted for 11 first-prizes and both grand championships. Two weeks later, against many of the breed's best, in a Register of Merit show at the Nebraska State Fair in Lincoln, they had the grand champion bull and four blue-ribbon winners. The following winter in Register of Merit competition at the Houston Livestock Show in Houston, Texas, the Wiese firm exhibited the best 10-head group, won the reserve bull championship, and five first prizes in class. A few years later at Denver, the Wieses had the reserve champion bull, best three bulls, two bulls and 10-head group.

Their Battle Intense bulls brought upper-bracket prices when offered in such consignment sales as that at Denver. Many went at long prices, ranging up to the $22,000 top figure of the 1965 event which was bid for their reserve champion of that year's show. The size, weight, bone and substance of the Wiese entries thus proved highly popular.

Further, and perhaps even more significantly, the descendants of Wiese-bred bulls exhibited by Wiese customers carried the Battle Intense banner with equal success. Such an instance, for example, occurred at the 1968 National Western when a bull carrying a Battle Intense top line won the grand championship, marking his fourth consecutive Register of Merit top purple.

As participants in the Iowa Beef Improvement Association and the American Hereford Association's Total Performance Records program, they also gave due attention to gaining ability and to feeding efficiency. By the early '70's, calf weights at 205 days of age were ranging up to 630 pounds and above, yearling bulls were approaching 1,200 pounds and three-year-old weights ranged up to 1,920 pounds.

A new ingredient was added to the Wiese breeding program in the early 1970's. This was the influence of the 2,685-pound Big Northern, a Canadian-bred bull owned by Glenkirk Farms of Maysville, Missouri. A double-bred grandson of this bull, Grand Slam, bred at Glenkirk and owned by Glenkirk and Wiese & Sons, not only won the Denver bull show's grand championship in 1976 but then went on to top the sale there when a half-interest in him was sold for $100,000. To that time, he was the heaviest summer yearling ever shown and sold at the National Western. Since then, the Big Northern family and some Line 1 influence have been combined with the continuing Battle Intense breeding in carrying forward the Wiese & Sons Hereford-breeding program.

Lester Wiese's innate ability to perceive what escapes the eye of others when looking at an individual or group of Herefords is a unique talent. How else can it be explained that at public auction he bought two breed-building herd bulls for $515 and $355, respectively, and paid $550 for the sire of another? These three, as cited earlier, were Anxiety Domino, Intense Domino and Battle Spartan 8th, the latter the sire of Battle Intense.

The Wiese breeding herd for many years consisted of from 75 to 100 cows, but as the decade of the '70's ended it had grown to well above 400 head. In fact, in the 1979-80 fiscal year of the American Hereford Association, only 21 herds recorded more calves than the 414 head listed for the Wiese firm.

Through most of the herd's existence, the Wiese custom has been to market

*One of Iowa's largest and best stock farms is that of Wiese & Sons, located just east of the central-west Iowa town of Manning. The farm home stood amidst the trees immediately to the left of the big silo in this view of the Wiese headquarters.*

the bulk of its output by private treaty, frequently in groups of substantial numbers to large-scale operators. A few have been consigned to such major sales, of course, as those at Denver, and sometimes to the state sale, but there has not been a Wiese auction since 1923, a testimonial in itself to the breadth and strength of the private treaty market for the Wiese product. Not only has this demand been nation-wide in scope, as shown by the fact that sales to buyers in 44 states have been reported, but international to the extent that Wiese Herefords also have gone to several other countries.

Les Wiese has been a firm supporter of organized Hereford activity throughout his adult lifetime. As early as 1921 he was serving as secretary of the Iowa Hereford Cattle Breeders' Association which held annual sales at Sioux City. He served as president of the West Central Iowa Hereford Association. He joined the Iowa Beef Producers Association in the 1920's and long served as one of its directors. He was one of the founders of the Iowa Hereford Association, was a long-time member of its board of directors and served as its president for four years.

For his contributions and accomplishments, Lester Wiese has received numerous honors. He was made an honorary member of the Block and Bridle Club at Iowa State College [now University] in 1930. The Iowa Beef Producers Association named him to its Hall of Fame in 1967. Special honors were accorded him at the Iowa State Fair's Register of Merit show in 1976 when he was hailed for "over 60 years of influence, direction and leadership" in the state's Hereford industry.

This then is the story of a man whose life has never been far from the cattle business, who was there when the family registered Hereford herd was started, and who revels not only in the contacts of the present but in his memories of the canny Scottish herdsmen and others whose acquaintance he made in times gone by. He accompanied the first Wiese show string to the Interstate Livestock Show at Sioux City, Iowa, in 1918, and to the Iowa State Fair in Des Moines the following year. He recalls vividly his first trip to Kansas City's American Royal in 1917, when it was held not at the stock yards as had been customary but at Electric Park, then a city amusement center.

The young man from Iowa, not yet 20 years old, looked, listened and learned as he stood at ringside. And his determination then and there was riveted toward a lifetime of endeavor in striving toward the production of Herefords as good as any to be found anywhere.

It is fair to say that from his beginning days down to the 1980's Lester Wiese never regretted for a moment his decision to make a lifetime career for himself as a breeder of Hereford cattle. And the breed was the beneficiary of that decision.

# Fred C. DeBerard

KREMMLING, COLORADO
1878-1956

PROBABLY no man's claim to fame in the Hereford industry ever was more broadly based than that of Fred C. DeBerard of Kremmling, Colorado. Beginning with a small commercial operation, which ultimately commanded national, and even international, attention, he subsequently established a registered Hereford herd. This was mainly because he could not find elsewhere bulls of the kind which completely filled his exacting requirements for use in his production of feeder calves and yearlings. In other words, the purebred herd was started for the purpose of supplying bulls for his commercial herd's needs.

It is entirely correct to state that the DeBerard record in the Hereford field thus was based upon three lines of successful endeavor: 1, cattle feeders who profited through the performance of DeBerard-bred calves in their feedlots and at the markets; 2, commercial ranchers whose herds were benefitted through the use of DeBerard-bred registered Hereford bulls, and 3, registered Hereford herd owners whose own breeding programs were advanced to a marked degree through the service of DeBerard-bred sires. Remarkably successful though he was in both commercial and purebred Hereford production, Fred DeBerard, with an abundance of laurels to look back upon, preferred, rather, throughout a fruitful lifetime to concentrate his thoughts and plans on goals not yet attained–on what might yet be accomplished.

Even had he chosen to look to the past, he could not possibly have perceived that his most lasting claim to fame as the 20th century winds down would rest upon two bull calves born a quarter of a century before his death. These were Advance Domino 20th and Advance Domino 54th, both bred by him, sired by his Advance Domino 13th, he by Advance Domino by Mousel Bros.' Prince Domino Mischief, and sold by him in 1933 to go to Montana where they became the key elements in founding the Line 1 Herefords which rose by the 1970's to a dominant position in the American Hereford industry.

Although Colorado was the locale for all of his Hereford accomplishments, Fred DeBerard was born on a farm near Cedar Rapids, Iowa, in a region long noted for its production of prime beef. His father, of partly French parentage, was a cattle feeder in conjunction with his farming activities, and a breeder of Durham cattle, as Shorthorns then were commonly known. Young Fred participated in the cattle-feeding work there, and later when the family moved to a place near the Illinois-Wisconsin line. Thus began the development of a set of principles which seemed always to guide his mature efforts: to produce the most high-priced meat with the least waste in an animal that would finish quickly and economically. He once was quoted as saying that he "never did care for an animal that carried a lot of cheap meat."

As he approached maturity, Fred seemed likely to follow in his father's footsteps. When he was 18 years old, his father purchased a prize-winning Durham bull out of the show herd of a successful Wisconsin breeder. That breeder had paid $1,000 for the bull as a calf in Canada. Calves by this sire, in the same DeBerard lot with calves of nondescript breeding, put on flesh more rapidly and finished out much better, the new owner's young son noticed. This was Fred DeBerard's first lesson in the value of good breeding. The second lesson came at about the same time, as he observed also that packer buyers liked those particular steers and were willing to pay much more for them. He saw at firsthand how important it was to produce the kind of cattle that the packers preferred.

The same year that his father bought the Durham show bull, Fred entered into the purebred business with the purchase of two registered Durham heifers. By the age of 23 he had a small herd of his own. But the west beckoned, partly because he had been told that it would be advantageous to him from a health standpoint. So he sold his Durhams, married a neighboring farm girl of whom he said years later that he "would have gone broke a long time ago if it hadn't been for her," and settled first near Julesburg, Colorado.

He noted that Herefords were the popular breed thereabouts, so he followed suit. But the going was hard, and also illness struck the young wife.

*Fame first came to Fred DeBerard through the championship performance of his carloads of Hereford feeder calves exhibited at such heralded events as those at Omaha, Chicago, Kansas City and Denver. In this picture he was displaying a load in the Denver yards during the 1949 National Western show.—Corwin photo.*

These factors, along with drouth and the fences that restricted a range operation, led to a decision to move again. Remembering that he had once seen shipments of top grass-fat cows on the Omaha market which, he was told, had come from the Middle Park and North Park regions of Colorado, he made up his mind to head in that direction. "If they can fatten cattle like that on grass alone, that's where I want to go," he said.

So in April, 1913, Fred and Mrs. DeBerard journeyed to Kremmling and purchased, with the aid of a government loan, the ranch in the shadow of Whiteley's Peak which became the home ranch of the DeBerard domain that eventually came to include seven properties. The "Peak" ranch, as it became widely known, was where Fred DeBerard really started his long and spectacular career with Herefords.

In assembling a herd of commercial Hereford cows, he picked them almost entirely for their apparent beef-producing qualities. He found such cows hard to buy, his best acquisition being a string of purebred but unregistered heifers which he got from the Burns Hole district in far northwestern Colorado in 1915. They were turned out with the grade cattle, and became the actual foundation of the great DeBerard commercial Hereford herd of later times. Scoffers thought he had paid too much for them, but the correctness of his determination to lay a quality foundation was borne out by later events.

DeBerard's last major purchase of commercial cows was made in 1929, and included some 700 head from the Jones Cattle Company of Kremmling. The buyer's insistence upon quality again was manifested in a unique manner. The 700–all regarded as good cattle–were divided into two groups, the better half being added to the DeBerard breeding herd and the lower-quality half going to market, which was the destination of any cows from his already established herd that did not quite measure up to the owner's standard. At the same time he bought the Jones ranch, going in debt $160,000 to make the deal. He later said, "It was good Herefords that pulled me through."

Fred DeBerard's nation-wide reputation was first established as an exhibitor of feeder cattle. His aim was not only to win but also to compare the product of his breeding program with that of other ranchmen and to test the acceptance of his cattle among cattle feeders. He began in the latter 1920's with yearlings and two-year-olds. He won the reserve grand championship on a load of yearlings at the first Ak-Sar-Ben Livestock Show at Omaha in 1928, but lost the grand championship to the calves of Dan D. Casement of Manhattan, Kansas. He had the first-prize load of Hereford yearlings at Denver in 1929, but calves of other ranchers won both purple ribbons. At the 1929 Ak-Sar-Ben he showed the winning load of two-year-olds as well as the first-prize calves, with the latter winning the reserve grand championship. His entries ranked second in both the two-year-old and yearling carloads at Denver in 1930. But calves won the top purples.

By now, DeBerard was convinced that the older steers, no matter how good, did not carry the bloom and eye-appeal to capture grand championships. While continuing for a time to exhibit yearlings, he decided thenceforth to concentrate his effort on calves. With them success came almost immediately as he captured his first grand championship on a load of calves at Omaha's 1930 Ak-Sar-Ben. In the same show, DeBerard yearlings placed first and second in class.

From that point on, DeBerard carloads of feeder calves established an unmatched record in carlot feeder cattle competition, winning more grand championships in major shows over a period of years than all other feeder cattle exhibitors combined. The show of shows of its era was the Golden Anniversary event staged at Kansas City's 1932 American Royal to mark the 50th anniversary of the American Hereford Association. A rich premium list attracted 88 carloads of feeder cattle which represented many of the West's "reputation" commercial herds. When it was over the judge, John P. Lynn of Tarkio, Missouri, had picked the DeBerard yearling load to head a class of 36 competitors, and followed suit in awarding the blue ribbon in the class of 52 carloads of Hereford feeder calves to the same exhibitor. And then he awarded the grand championship to the calves. The caliber of the competition is indicated by the fact that entries from the famous Matador Land and Cattle Company's Texas ranches ranked second in each of these classes, with other nationally known names on down the list. In DeBerard's first Royal show, he came, saw and conquered.

That performance was by way of a prelude to the following year. In 1933 DeBerard calves won feeder carload grand championships consecutively at the four major events of their kind–Omaha, Kansas City, Chicago and then, in January, 1934, at Denver. This truly sensational performance was subsequently duplicated in four other years.

DeBerard emphasized the remarkably uniform high quality of his feeder cattle output in an advertisement in *The American Hereford Journal* following the 1933-34 season. In it he stated that all of his 1933 steer calf crop old enough to wean had been disposed of at those four show points–not only the winning loads but also some 200 other calves. Such was the demand engendered by the performance of the herd's output. "These calves were all from my grade cows," DeBerard later said, "and sired by registered bulls of my own raising."

*Fred DeBerard on the job with a pen of bulls on his home ranch in Colorado's Middle Park. This picture was taken in the 1930's shortly after his purebred herd began to gain wide prominence.–* Hereford Journal *photo.*

In all, the records show that he bred and exhibited 50 grand champion loads of feeder calves, including seven in a period of 11 years at the once-heralded Chicago Feeder Show.

Years of effort, of course, were back of these achievements, beginning with foundation cows of the type and quality he demanded. Then came the search for bulls. He did not consider the commonly termed "range bulls" good enough, and frequently competed with purebred breeders for the better kinds. But even then, the ones he desired were scarce, if they could be found at all. This was what led him into breeding his own. His purebred herd was established in 1918 with the purchase of 16 cows from Jack Dickens of Walden, Colorado, at $500 each. These featured the bloodlines of the herd of A. B. Cook of Townsend, Montana, noted for its Beau Brummel-Panama line, from which had come the Denver show's grand champion bulls in 1914 and 1917, and the senior and/or junior champions in 1916 and 1918. Other females came later to the DeBerard ranch, including nine bought in the 1923 Denver sale. By 1934 his advertisements in *The American Hereford Journal* indicated that his purebred breeding herd consisted of some 200 head. After 1920, virtually all bulls used to sire the DeBerard prize-winning feeder cattle came from the owner's registered herd.

While this program solved the bull problem in his commercial herd, the need for ever-better bulls for the registered herd was intensified. Prince Domino bloodlines eventually predominated in the development of the registered herd, although his emphasis seemed mainly to be on the type and conformation that met his ideal rather than on pedigree. However, he obviously recognized, too, the importance of the latter. His first highly successful cross was made with Prince Domino 130th by Prince Domino on cows of Regulator breeding, descended from the Repeater line of O. Harris & Sons of Missouri, plus others of Beau Carlos descent. That resultant herd, plus infusions of the blood of Onward Domino by Prince Domino, through two sons, was the breeding primarily back of those early-day feeder calf winnings.

Later came sons of Advance Domino, a grandson of Prince Domino, including Advance Domino 13th for which DeBerard paid $1,000 as a calf. Other bulls of Hazlett, Chandler and Wyoming Hereford Ranch production also were acquired during this same general period. This program resulted in cow herds, both registered and commercial, which carried much of the most popular breeding of the times, although the owner's ideal as to individuality seemingly continued to be the prime consideration in selections and matings.

While building up his registered herd, DeBerard found neighbors desiring to buy bulls. Many wanted to obtain the bulls that he had used in the grade herd, and he built up quite a trade on these three- and four-year-olds. The fame of these bulls spread and North Park cattlemen started buying heavily of them from him. Then commercial ranchers from farther points who had observed his results at the stock shows began to seek bulls from the herd, including such industry leaders as the Monahan Cattle Company of Hyannis, in the Nebraska Sandhills. This firm bought repeatedly and then sold the resulting Circle Dot feeder cattle to Fred Attebery of Mitchell, Nebraska, who time after time topped the Chicago market with his carload shipments of finished beeves.

This, however, was not the first occasion when steers carrying DeBerard breeding commanded industry attention in Chicago. As far back as 1931 Fred DeBerard sold his third-prize carload of feeder calves at Denver to John D. Moeller of Schleswig, Iowa, who fed them out and exhibited them to the grand championship at that year's Chicago International carlot fat cattle show. Other DeBerard-bred loads fed by others won blue ribbons also at this then-greatest of fat cattle shows, accounting in 1938, for example, for two firsts and two seconds in the three Hereford classes.

Although never extensively, DeBerard occasionally exhibited breeding cattle, especially bulls, at Denver's National Western, where his bulls ranked as high as reserve champion on two occasions. In 1940, he won the coveted purple in the carlot bull division, thereby breaking a string of five consecutive grand championships captured by Wyoming Hereford Ranch. This and other DeBerard successes, both in the registered and commercial fields, led to the development of a market for DeBerard registered Herefords in many sections of the United States, and in Canada, Mexico and South America.

It is appropriate here not just to state but to emphasize that the ultimate accolade came when the descendants of the DeBerard-bred Advance Domino 20th and Advance Domino 54th, sons of his Advance Domino 13th, became a dominant element in the registered Hereford industry during the 1970's. The former, winner of first prize in the senior bull calf class of 14 head at the 1933 Denver show, tied another entry for the top bull price of $500 in that year's National Western sale in going to the U.S. Range Livestock Experiment Station at Miles City, Montana. There the blood of these two was intensified in a unique breeding program covering a period of many years in the development of the Hereford family which came to be internationally known as the Line 1's. Few major herds by 1980 had not felt its influence: A belated tribute, indeed, to the taciturn "old master" with the steel-grey hair, square jaw and closely trimmed mustache who said it all when he once stated that "chance did not build my herd."

# CHARLES H. HARRIS

FORT WORTH, TEXAS
1871-1958

DR. CHARLES H. HARRIS whose constructive effort brought wide prominence to his Harrisdale Hereford Farms near Fort Worth, Texas, especially during the second quarter of the 20th century, had unusual opportunities to gain insight into genetic engineering long before such terminology came into more general usage. As a student of medicine both in the United States and Europe, he continued, even after years of practice as a leading physician and surgeon in Fort Worth, to keep abreast of developments relating to health and well-being, and became intrigued with the role of heredity.

On one occasion he pointed out that mental health centers, jails and penitentiaries confine many boys who have good fathers, yet it is rare, he went on, "that such boys have both good fathers and good mothers." Dr. Harris added: "The highest type of civilization comes from the mating of the best fathers and the best mothers. This holds good in all animal reproduction. Chromosome cells are chemical cells. They do not know bloodlines but reproduce according to the character of the chromosome cells themselves." Consequently, the doctor concluded, the present and past performance of the best sires and the best dams should guide the constructive breeder's effort toward ever higher attainment.

The implementation of this guiding principle led to the production in the

Harrisdale herd of one of the greatest herd bulls of his time, and by all odds the key bull in making that herd one of the breed's genuine history-makers. That bull was Prince Domino Return.

But before delving more deeply into this noted bull and how he came about, it might be well to go back to the beginnings of Dr. Harris as a Hereford breeder. From humble origins in his native Texas, he pursued frugal practices in order to graduate in 1894 from medical school, whereupon he first entered into the medical profession at the small West Texas town of Moran. From the outset the able young doctor was successful, and ambition led him in due course to the city of Fort Worth where he proceeded to carve out a truly distinguished career in his medical practice.

But as with numerous other surgeons who find surcease from the strain and tension of their profession in rural surroundings, and especially in working with animals, Dr. Harris in 1916 bought his first Herefords, more or less with the idea of running a small herd as a hobby. As a youth when riding the range

*Focal point here is Prince Domino Premier, 13 times a blue ribbon winner and five times winner of the purple in 1937 and 1938 as a member of the show herd of Harrisdale Farms, the latter triumphs including the championship award at the 1938 National Western Stock Show. He was pictured here on February 7, 1939, just after having been sold for a sale-topping $8,000 to Joseph W. Radotinsky of Kansas City, Kansas, left, and Warren V. Woody of Barnard, Kansas, second from left. The seller, Dr. Harris, stood next to the bull on the right, along with Col. Earl Gartin, the auctioneer, from Greensburg, Indiana.*

he had made his first acquaintance with Herefords, but an urge to practice medicine had led him then to lay aside his boots and spurs.

By the time his interest in cattle reasserted itself, his practice had prospered sufficiently for him to acquire land and buy a set of Hereford cows from the Hereford Grove Ranch in Childress County, Texas. Appropriately enough, he later pointed out, their sire was a bull named Mortgage Lifter.

Then, as his interest heightened, he bought from B. N. Aycock & Son of Midland, Texas, a number of daughters of the 1917 Fort Worth show's grand champion bull, Hector, he a descendant of Beau Brummel. The first Harris herd bull of note was Prince Domino 76th, one of the earlier sons of the famous Prince Domino, bred by Otto Fulscher of Holyoke, Colorado, and out of a daughter of Beau Aster by Beau Mischief. This bull thus exemplified the Prince Domino-Beau Aster cross, one of the most successful in breed history.

As a student of inheritance, Dr. Harris by now had decided to concentrate his efforts on Herefords of Gudgell & Simpson ancestry. And he set about doing it by adding a substantial number of females representing some of the best of the Gudgell & Simpson lines. Some 15 head, including several daughters of the champion Beau Randolph, a G. & S.-bred son of Beau President, came from the herd of H. Gaudreault & Son of Hastings, Nebraska. By the latter 1920's he was buying at various places upper-bracket females, including a Prince Domino-Beau Blanchard cow at $3,050, and a Prince Domino daughter for $1,900.

In the dispersion sale of the noted herd of Jesse Engle & Sons of Sheridan, Missouri, in January, 1927, he selected a dozen of the herd's top females, including five daughters of the famous Beau Blanchard and four sired by one of that sire's top sons, Beau Blanchard 95th, co-top-selling bull in that auction. He bought females in the dispersal of the prominent Anxiety 4th-bred herd of P. J. Sullivan of Wray, Colorado, and was the leading purchaser of top-selling cows in the dispersion of the well-known Anxiety 4th herd which had been built by C. O. Keiser of Canyon, Texas, getting seven at an average of $1,236, a figure more than three times that which was recorded on all of the females in that auction. This group included three daughters of Prince Domino. About the same time he got 10 daughters of the Anxiety 4th-bred Don Stanway from B. H. Conner of Claude, Texas. And he bought others elsewhere.

All the while Dr. Harris was seeking just the right herd bull for mating with these richly bred females. And he found him late in 1928 in the herd of W. T. Womble & Sons of Hereford, Texas. In fact, Dr. Harris had been well aware of the bull earlier, but until now could not persuade the Wombles to sell The Prince Domino to him. The price of $10,000 established a new high for the decade of the 1920's. Sired by Fulscher & Kepler's Prince Domino and out of

*Dr. Charles H. Harris displayed here not only a trio of Harrisdale heifers in their show stalls but also a silver trophy which was captured by his entries. The herd's representatives competed widely for many years.—Phil Winegar photo.*

a dam by Young Anxiety 4th by Bright Stanway, The Prince Domino was bred by Mousel Bros. of Cambridge, Nebraska, who had sent the cow to the Fulscher & Kepler ranch to be mated with their landmark herd bull. The Wombles bought him in 1926 for $2,500, and only the persistence of Dr. Harris persuaded them to let the bull go.

Although Dr. Harris already owned a cow herd of sterling individuality and royal ancestry, for a special mating with the new bull he continued to bear in mind the early admonition regarding inheritance—the essentiality of the very best possible mother if the ultimate result is to be achieved. In the so-called "criterion sale" held March 12, 1930, in connection with the Southwestern Exposition and Fat Stock Show at Fort Worth, he found exactly the heifer which fulfilled his dream. She was the exceptional summer yearling Blanche Mischief 14th, richly bred in the best Anxiety 4th tradition. To top it off, she had been acclaimed junior and grand champion female of this great show, a remarkable feat for a heifer only 20 months old.

When Dr. Harris made the final bid of $4,200 for the heifer, at a time of wide financial depression, it is probable that all of the persons in that large audience who believed that he had made a deliberate investment upon which he could hope to obtain substantial returns could have been counted on the fingers of one hand. But he was undeterred. This was the heifer which he had envisioned. When Prof. H. J. Gramlich of the University of Nebraska awarded her the supreme purple not a single dissent was heard, according to the show report in *The American Hereford Journal.* In fact, she had been pronounced by many top breeders and good judges as being as nearly perfect in conformation and type as any female they had ever seen. Dr. Harris certainly concurred. Blanche Mischief 14th approximated his ideal. It was easy for him to anticipate that her mating with The Prince Domino would yield the latter's successor at Harrisdale.

Further, he liked the similarity in bloodlines. It had been fully demonstrated that the blood of Prince Domino and that of Mousel Bros.' Beau Mischief was congenial and that their mingling produced especially good results. Blanche Mischief 14th carried a high percentage of the blood of both of these great bulls. She was linebred to Prince Domino Mischief, rated by most experts as the greatest breeding son of Prince Domino, while Prince Domino Mischief's dam was a daughter of Beau Mischief. The heifer's sire was Advance Mischief Jr. by Advance Mischief, one of the most noted sons of Prince Domino Mischief, and her dam was by Prince Domino Mischief.

Dr. Harris never entered into any other business transaction more deliberately than he did in the purchase of this grand champion heifer. If any factor other than his dream of the "perfect" mating entered into his decision, it was sentiment. Dr. Harris as a young physician practicing at Moran, Texas, had been in charge at the ushering into this world of Bertram A. Elliott, junior member in charge of the firm of R. A. & B. A. Elliott, owners of Elliott Hereford Farms, where this superb heifer was bred and developed. The cycle had come full circle, in a sense.

Dr. Harris had a bit of bad luck with his heifer. He got only one calf from her. But he also had some good luck. From her first and only mating with The Prince Domino, she produced the calf which gained international fame as Prince Domino Return. He was the key sire in what was achieved at Harrisdale during the some 40 years of the herd's existence. His get included many top show cattle–winners of high honors in national competition. In a two-year span during the latter 1930's Harrisdale entries won hundreds of ribbons at state fairs and the major stock shows. These winnings included the "best 10 head" at the 1937 Chicago International and at the 1938 American Royal show in Kansas City. One of his sons, Prince Domino Paladin–whose dam was by the first significant Harrisdale herd bull, Prince Domino 76th–was grand champion bull in major competition during the 1937-38 season. Another son, Prince Domino Premier, was 1938 Denver champion. Other sons and daughters also won heavily.

Prince Domino Return's influence was widely disseminated through Harrisdale auction sales—two a year over a considerable period of time. At the peak of his usefulness in the middle 1930's he was serving up to 150 cows a year, which made large numbers of his get available. Many of his sons contributed substantially in passing along their sire's influence to later generations of Herefords in leading herds across the country. One of those sons became champion bull at the 1939 National Hereford Show at Dallas, Texas, for Dean Ranch of Fort Worth, Texas, and another gained recognition as the grandsire of the 1946 Fort Worth show's champion bull exhibited by George D. Keith & Son of Wichita Falls, Texas.

Obviously in a breeding herd of such substantial size, many Harrisdale-bred bulls were sold for commercial range-herd duties. One of the major buyers of Prince Domino Return sons for range service was John C. Burns, manager of the extensive Texas ranches of the Estates of S. B. Burnett and Tom L. Burnett, with office headquarters at Fort Worth. Long a leader in the beef-cattle industry of the Southwest, Burns displayed his approval of the sons of Prince Domino Return by his extensive and repeated purchases of them at Harrisdale and he once commended the old bull himself for "his great head, good front and rear quarters, level top, good bone and straight, well-set legs." Burns added that the bull was "quite uniform in his width from end to end, smooth fleshed, mellow hided and full of quality." If he were to be criticized at all, Burns continued, "it would be mainly for being a little too long." Perhaps he was a bull ahead of his time!

Not only did Dr. Harris carve a lasting niche as a Hereford breeder, but also as an eminent physician and surgeon and as an astute hospital executive. His Harris Clinic-Hospital and the Harris School of Nursing were his own private institutions. When the Methodist Hospital in Fort Worth, long the major medical center in the city, encountered financial difficulties in 1937, Dr. Harris came to its rescue and by means of large monetary assistance coupled with his own effort and experience brought it through the crisis. Upon his retirement from active practice the institution became Harris Memorial Hospital. Through the Charles H. Harris Foundation, Dr. Harris in 1943 also provided the means to assure the continuation of its college of nursing.

As age crept up on the good doctor, he gradually reduced his cattle holdings, a dispersal at auction being held in June, 1949, and the final group being closed out at private treaty in 1954. Four years later he was gone, at the age of 87. By then he was entitled to look back with considerable satisfaction upon two careers in which he had attained uncommon success.

# Robert W. Lazear

CHEYENNE, WYOMING
1889-1957

ROBERT W. LAZEAR was one of the most remarkable men ever associated with the Hereford industry. First of all, he was a city boy, born and reared in Chicago. He became an engineer, receiving his bachelor of engineering degree from one of America's most prestigious educational institutions in that field, the University of Michigan at Ann Arbor. He began to practice his profession by undertaking a substantial irrigation project on Colorado's Western Slope, during the course of which he took several shipments of Holstein cattle and Duroc hogs to the area. While engaged in this work, he became enthused over the contribution which he believed improved livestock could make to the agricultural advancement of this, or any, community.

When the irrigation project proved to be an impossible undertaking, the owner of the property, Henry P. Crowell of Chicago, had another possibility in mind for the still youthful engineer. Crowell had loaned some money on a ranch near Cheyenne, Wyoming, owned by the Hereford Corporation of Wyoming. When financial difficulties developed there in the depression period immediately after World War I, and the owners decided to opt out, he apparently had no recourse but to take over the property. Suddenly he found himself to be the owner of many thousands of acres of ranch land and several hundred head of good registered Herefords.

Obviously an operation of this scale required strong management if the new owner's investment was to be recouped. Thoughts that were turned over in Crowell's mind as he considered the possibilities were once spoken: "Should the new manager be young, middle-aged, or old? Should he be thoroughly imbued with the cattle industry? Or, young, inexperienced but capable, an independent thinker, courageous, and ready to find new answers?"

Then Crowell gave his decision in these revealing words: "We decided in favor of the young man. We believed more could be accomplished by him than through someone set in his ways of thinking and fixed in his judgments."

Thus it was that this strong, active, virile man, whose keen-eyed, trim-figured alertness became almost a personal trademark, came into the picture at the Cheyenne ranch, and figuratively grew in stature to match it. In later years the mention of one almost anywhere in the beef-cattle world came also to infer the other. It took some deliberation before Bob Lazear decided to leave the profession for which he had been trained and step into something entirely new. But when he did accept, early in 1920, he entered into the task with a will. He first carried the old firm through receivership, and then set up the Wyoming Hereford Ranch as the new company which assumed ownership of the land and cattle in August, 1921.

With steadfast purpose, he first surrounded himself with assistants of known abilities. Then with the methodical logic of an engineer he set about to learn all he could about Herefords and the ranching business. Always with high ideals, plus tenacity and a tremendous will to work hard, he developed the keen business sense and know-how that carried him to the top echelon of the Hereford industry.

Lazear was an enterprising man, an innovator and a do'er. Within a remarkably short time, his touch became apparent in every phase of WHR operations. Not only did his managerial skills encompass the handling of the 60,000-acre property and its 2,000 or more registered Herefords, but he also was constantly aware of the

*Robert W. Lazear, who as a young man in the early 1920's assumed the management of Wyoming Hereford Ranch near Cheyenne, was charged with this responsibility by Henry P. Crowell of Chicago, right, who had just become owner of this famous property and one of the industry's largest registered Hereford herds.*

*No other exhibitor ever came close to winning as many grand championships in carlot bull competition as did Wyoming Hereford Ranch at Denver's National Western Stock Show, the greatest of all such events, between the middle 1920's and the middle 1950's. This was WHR's grand champion load at the 1954 show.*

details incident to handling both the herd and the ranch. This perhaps was a reflection of his training and practice as an engineer, a field in which recognition of the vital nature of details is of paramount importance.

Throughout his career in the Hereford field, Lazear was quick to give credit where he believed it was due. A few years after he assumed the WHR helm, a Colorado neighbor, Otto Fulscher of Holyoke, became associated with WHR for some four or five years, bringing with him a select group of cows and the famous herd bull, Prince Domino, which lived his final years at WHR.

This Lazear characteristic once was voiced in these words: "I should like to pay tribute to the most able tutelage of that veteran master breeder, Otto Fulscher, with whom it was my privilege to be associated. He exemplifies in the highest degree the source of information, wisdom and power upon which every Hereford man should draw in rounding out his own interest in and quest for knowledge in Hereford cattle."

Whereas bulls of several lines, including Hazlett, Fairfax, Beau Donald, Beau Blanchard and Bonnie Lad were in service in the herd when Lazear became manager, he had added a son of Prince Domino to the sire battery even before the Fulscher connection was accomplished. Thereafter, Prince Domino sons and their descendants were dominant factors in the herd for at least a third of a century, and through it contributed greatly to the family's fame and that of WHR itself.

To a greater degree perhaps than any Hereford breeder of his time, Lazear became an exponent of personal salesmanship. He attended literally hundreds of auction sales from coast to coast in which Herefords of WHR extraction were being sold to assist the sellers in every way possible. Likewise, whether or not WHR was an exhibitor, he made it a practice to be at the ringside when WHR customers were displaying their product in order to lend such support as might be in order. It is entirely possible that he was "on the road" in behalf of WHR's interests for a greater proportion of his time than any of his contemporaries, and there is no doubt that his friendly personality and easy informality yielded good dividends.

With his time seemingly so fully occupied, it is little less than amazing to learn that the WHR herd records, advertising programs and correspondence also were very much his personal domain. It is probable that his imaginative skill and talent for advertising layout and copywriting have never been matched by any other stockman–perhaps again reflecting his engineering background.

He personally prepared virtually all of Wyoming Hereford Ranch's advertisements during the 35 years of his tenure, and the WHR displays in all publications bore the unique trademark of his talent. Literally scores of unusual direct-mail brochures and booklets were sent to the ranch's customers and prospective customers through the years, which served as constant reminders of WHR and its Herefords. For several years he issued from the ranch an occasional small publication, *Brands*, which was counted on to keep the ranch in contact with the industry, this, too, reflecting the Lazear touch.

The question becomes: how well did the WHR Herefords perform under

*Bob Lazear, at right, here was showing a couple of senior bull calves by the imported Vern Diamond to visitors, from left, W. A. Crawford-Frost of Caerleon Ranch, at Nanton, Alberta; J. B. Cross of Calgary, Alberta, owner of Bar Pipe Farms at Okotoks, Alberta, and A. V. Harrell of Ellensburg, Washington.*

*Great crowds from all points of the compass thronged the ranch headquarters and overflowed this sale pavilion for the annual WHR auctions held each October for many years.*

the direction of this man who learned his job while doing it? The answer is: remarkably well. Beginning in 1933, auction sales which reflected the great demand that had been cultivated for the WHR output were held on a regular basis at the ranch. The climax came in the 1947 auction when 72 head commanded an average of $5,934, with top individual prices of $61,000 and $53,000. Both the average and the tops established new breed records to that date.

Show herds were sent to the leading fairs and expositions from Chicago to the West Coast and literally hundreds of blue and purple ribbons were awarded to WHR bulls and heifers. Few if any of the cattle exhibited were not bred at the ranch, the show strings thus serving as advertisements of the breeding herd's prowess in producing top-quality Whitefaces.

Lazear, however, recognized that commercial range cattle producers comprised the primary customers for WHR's volume bull output, an important point for a herd which in some years recorded more calves with the American Hereford Association than any other in America—occasionally well over 1,000 head. The WHR herd, in fact, was operated mainly on a range basis, a fact which made a favorable impression on many good commercial herd owners who became steady WHR customers.

Wyoming Hereford Ranch, during Lazear's tenure, became the most successful exhibitor by far in the history of the greatest of all carlot bull shows—the National Western at Denver. The first WHR grand championship in this category was won in 1926, and thus commenced a 29-year span in which WHR bulls were Denver carlot grand champions 20 times. Lazear seemed to take greater pride in this record than in any other of WHR's accomplishments under his guidance.

This entirely unassuming man also developed skill as a judge of beef cattle, not only of Herefords but also of other breeds. He officiated at the great Commonwealth Hereford Show at Regina, Saskatchewan, in 1955, and in previous times had judged at the famous Calgary Spring Bull Show and also at the National Polled Hereford Show. He also became a leader in breed councils, serving three terms as a member of the board of directors of the American Hereford Association. For the last 20 years of his tenure as WHR manager, he also served as managing trustee of the Wyoming Hereford Ranch Trust. This was set up by Henry Crowell to insure the continuing operation of the ranch and herd for a fixed period, with the profit, after operating costs, to be distributed through an evangelical Christian trust.

A signal honor came to him in 1953 when the University of Wyoming at Laramie awarded him an honorary LLD degree. This is the University's highest award and is reserved for those who have distinguished themselves by their contributions to the advancement of their state and country and the welfare of their fellow citizens.

It is a tribute in itself that during his period as manager, Wyoming Hereford Ranch gained international stature as a tourist attraction. It was listed for years in the Tour Book of the American Automobile Association, and hundreds who stopped month by month at the village-like headquarters east of Cheyenne were met with cordiality whether or not they knew anything about Herefords and cattle ranching.

The ranch and its manager were the subject of feature articles in the heyday years of the *Saturday Evening Post* and *Country Gentlemen*, and also were featured in the *National Geographic.* A portion of an Arthur Godfrey radio broadcast once originated on the ranch, which long since had become an institution which reflected most favorably from the standpoint of the general public upon the Hereford cattle industry.

Lazear was blessed with an active mind. He always believed that there was much more he could learn, and he never stopped trying. When the recessive genetic factor, commonly known as dwarfism, began to manifest itself in the herd near the end of his tenure, he did not deplore his bad luck but merely buckled down with resolute courage and vigor, working perhaps more intensely than ever before to rid the herd of an insidious element that had been introduced unwittingly a half-century earlier, during a period of different ownership and management, but which had remained hidden for decades.

It was generally agreed that Bob Lazear was "quite a man," and quite a ranchman, and that his endeavors reflected creditably upon the Hereford industry in which he was involved for most of a lifetime.

# ARTHUR W. THOMPSON

LINCOLN, NEBRASKA
1887-1970

SELDOM in livestock history has an auctioneer become so proficient in his profession and so popular among his clients that he ultimately decided to devote his time and talent entirely to a single breed. So it was with Arthur W. Thompson of Lincoln, Nebraska, who gained fame in the middle years of the 20th century as the premier Hereford salesman of his era, and possibly unsurpassed in livestock auctioneering history.

He began life on the farm in York County, Nebraska, where his father had settled as a pioneer in 1865. By the age of 12 or 13 years, he was doing a man's work, as was customary in those times, but when the farm work and chores were finished he occasionally accompanied his father to farm auctions in the community. There was an attraction in this profession which he found intriguing, and he followed every move of the auctioneer, his sharp ears catching every word and every inflection, and his eyes taking in every detail. Before long, as he trudged along behind the walking plow or cornfield cultivator, with only the family team as listeners, he practiced his logic in behalf of improved livestock as a means to a more prosperous farm economy, and presented his version of the auctioneer's chant as he called for bids beneath the blue Nebraska skies.

Although he wasn't ready to make his move quite yet, young Thompson

*As the crowd assembled for an auction, Col. A. W. (Art) Thompson sometimes appeared in the auction box a little before sale time to level his gaze upon the ring and the gathering crowd as if to contemplate the outlook. This was one of those occasions.*—Hereford Journal *photo.*

had made up his mind by the time he was 15 or 16 that he would become an auctioneer. He once explained his decision in these words: "As I walked for miles down those long rows in the corn field under the broiling sun I was determined to get into a line of work where I could get a drink of cold water any time I wanted it." So when he was 19 years old, Art Thompson became an apprentice auctioneer, usually working with little or no pay, under the local farm auctioneer, Col. T. W. Smith.

But this does not indicate any neglect of his desire for education beyond that obtained in the country school in the family's farm neighborhood. He attended York College in his home county-seat town of York, and then went to Lincoln Business College in nearby Lincoln, Nebraska. And then he moved decisively in the direction of a full-time career in auctioneering by attending the Jones Auction School in Chicago. All the while he attended every possible sale, read the agricultural papers and livestock journals, and dreamed that some day he might follow in the footsteps of such noteworthy purebred auctioneers of the times as Col. F. M. Woods and Col. R. E. Edmonson.

When Art Thompson's local mentor, Colonel Smith, received a political appointment which necessitated a move to Lincoln, the budding young auctioneer fell heir to the Smith clientele, later recalling that he conducted something like 130 sales in his first year. This marked the actual beginning of a career that the young Nebraskan, even with all of his ambitious enthusiasm, could not even vaguely have foreseen. One of his first thrills, he once recounted, came when he sold a purebred sow in one of those early farm sales for $100, an almost unheard of price for that time and place.

In thinking back to that day, Thompson recalled that he visited the farm before sale day, studied the sow's conformation and quality, and talked with her owner regarding her history and pedigree in order to be ready for what he anticipated would be the high point of the day from his standpoint. What this tale does is highlight a Thompson characteristic which later was to contribute to his remarkable success in selling purebred livestock of all breeds–and finally Herefords exclusively. In essence, he studied to maximize his knowledge of what he was going to sell, he worked on the logic of his presentation, and laced it with his special brand of persuasiveness.

Thompson was 34 years old when he was first called on to conduct a registered Hereford sale. This was on December 6, 1921, and the sellers were H. Gaudreault & Son of Hastings, Nebraska. Col. Fred Reppert of Indiana, then the No. 1 Hereford salesman, was the scheduled auctioneer, but a sudden illness prevented him from appearing. Thompson was at home in York when a call came asking him to go to Hastings at once to handle the Gaudreault sale. It was only an hour or so until sale time, Hastings was 70 miles from York, and the road in those times was not very good. But he got there and thus marked the beginning of a career that was to span thousands of sales and incalculable miles of travel in criss-crossing North America hundreds of times.

For more than 20 years after that 1921 beginning, Thompson sold purebred livestock of all breeds, frequently maintaining a schedule which involved a sale every day for six days a week. Finally, in 1944, because of his own personal inclination toward the Whitefaces as well as to ease the pressure on himself, he resolved to confine his auction work exclusively to the selling of Herefords. And even then he could not keep up with all of the requests received for his services.

He was eloquent as he assumed command of proceedings in the auction box. His sale-opening comments frequently were classics in their organization and forcefulness. Not only was he an orator of no mean ability but he was convincing, logical and fully informed. He was an excellent businessman, one who knew beef cattle values and understood the cattle business. In sale after sale he cited persuasive examples to support his arguments in behalf of improved livestock and the soil conservation practices to which he believed animal agriculture contributed immeasurably. Many a man bought his first registered Herefords because of logical, practical, business-like statements he had heard made by this auction salesman who preached the gospel of improved livestock wherever he went. He was in complete agreement with an observation once made by John Lethan, first full-time field representative of *The American Hereford Journal*, that a properly conducted auction is “a businessmen’s convention from start to finish.”

Colonel Thompson also was an innovator. He firmly frowned on language and stories that might be considered offensive. In earlier times members of the fairer sex usually stayed away from salerings, but he urged their attendance. This frequently resulted in a prospective buyer making one more bid when

*Col. A. W. Thompson never missed an opportunity to widen his acquaintances and cultivate his friendships in Hereford circles. In the heavy winter coat which was somewhat of a seasonal trademark, he was seated here in an alley in the carlot bull division at Denver's National Western in 1939. Others, left to right, were: C. H. Andrews, Medford, Oregon; Walter E. Cole, Broken Bow, Nebraska, a load of whose yearling bulls had just placed second in the show; Thompson; Henry Volzke, Waco, Nebraska, and J. H. Melville, Broken Bow. When Cole's 500- to 600-head herd was dispersed two years later, Thompson was the auctioneer.*—Hereford Journal *photo.*

the wife indicated approval. He brought other departures to auction procedures and introduced such elements as loudspeaker systems which had not previously been used, at least extensively, in selling livestock.

Pages would be required to cite all of the historic Hereford auctions of which he was the chief auctioneer and all of the record-breaking prices which were established during the years until his retirement in 1953. He was the auctioneer in charge of virtually all of the leading sales of registered Herefords during the heyday of his career, and his "book was full" most of the time. In fact, it was so distressing to him to have to turn down clients who sought his services that finally he made an arrangement with an auction company through which his bookings were handled. Only by this means did he finally succeed in putting on the brakes.

It was in the January 1, 1922, issue that A. W. Thompson put his first modest advertisement in *The American Hereford Journal*, an advertisement one inch deep and one column in width. Alongside the picture of a good-looking young man was this:

"SELLING PUREBRED LIVESTOCK is my business. Years of experience enable me to give efficient and satisfying service. I respectfully solicit a share of your sales. Terms moderate. A. W. THOMPSON, YORK, NEB."

Within a year or so he moved to Lincoln, and from there began his climb up the ladder until finally in the latter 1940's he received a fee of $16,000 for conducting a single two-day sale. Many distinctions came his way, which he acknowledged with unassuming pride. He once pointed out that up until the time of his retirement it had been his good fortune to be the auctioneer who handled the actual selling of every Hereford that had been sold at public auction at prices of $50,000 and above. He also found satisfaction in the fact that he had been the auctioneer in charge of the only three Hereford sales up to that time that had grossed as much as a million dollars. His highest-averaging sale, at $5,934, established a new record for the breed that never before had been approached.

But it was another, earlier sale that Colonel Thompson once cited as perhaps his most exhilarating experience as an auctioneer. This was the June, 1937, dispersion sale of the great Hazford Place herd of the late Robert H. Hazlett at El Dorado, Kansas, one of the greatest assemblages of Herefords in the entire history of the breed. Under the magnetism of his salesmanship the 604-head herd of that master breeder brought a total of $305,250, for a $505 average, sensational figures for the depressed times then prevailing. Considering the inflationary prices of later periods, it has been speculated that the sale would have grossed from two to three million dollars had it been held, for example, during the "good times" of the 1970's. And that would have reflected an average of $4,000 to $5,000 a head.

This historic event was the one which cinched Colonel Thompson's rank in the auctioneering hierarchy. He occupied the box throughout the entire three days of the sale, the actual selling requiring a bit more than 20 hours. He held the close attention of the vast assemblage during the whole time and there was not a minute in which he did not have full command of the audience, according to reports at the time. His performance, too, was a demonstration of stamina that has never been matched in an auction arena, since he actually officiated on the block in

*A major high point in the career and life of Colonel Thompson came with the unveiling of this portrait of him in the famous gallery of the renowned Saddle and Sirloin Club in 1951.—Willming photo of Hoffler portrait.*

the selling of every animal that went through the ring in that historic event. From then forward there never was any doubt that he was the leading auctioneer of the times in the Hereford field.

Naturally many distinctions came to Colonel Thompson, but perhaps none pleased him more than when his portrait was unveiled in the gallery of the Block and Bridle Club at the University of Nebraska at Lincoln. Although he never attended the university, he was a firm supporter of it and its educational goals.

In 1951, a great assemblage of Colonel Thompson's friends from coast to coast gathered in one of the great halls of the lamented Saddle and Sirloin Club in Chicago during International Live Stock Exposition week to pay him honor. This great occasion marked the unveiling of his portrait in the Club's famed gallery of livestock industry leaders. He is the only American auctioneer ever so honored.

As a speaker at the Block and Bridle Club ceremony at Lincoln, Dan D. Casement, noted cattleman of Manhattan, Kansas, eloquently summed up the sentiments of thousands in these words:

"The imperative purpose which I had when I came up here was to tell this man, Art Thompson, how much I admire him. He is honest, he has integrity, and the livestock industry of America owes him a great debt of gratitude because of the splendid qualities he has brought into the business of distributing its product. It is a wholesome experience for me to greet him because I recognize in him so much sterling manhood, so much character, and the splendid human quality that draws people to him because he has the natural good manner and kindness that come from a great, kind heart."

# ROY R. LARGENT

MERKEL, TEXAS

1903-1968

ROY R. LARGENT was born into the Hereford business and lived with it every day of his 65-year lifetime. His father, C. M. Largent of Merkel, Texas, had left his drugstore operation at McKinney, Texas, to buy a ranch at Merkel, and started with Herefords at that location in 1900, more than two years before his son, Roy, was born.

The youngster lost no time in getting right into the midst of the Hereford action. He accompanied his parents to his first Fort Worth Livestock Show, as it was then known, in 1904, and was back again for the 1905 and 1906 shows. They missed the 1907 event, however, because of the imminent arrival of a new member of the family, C. M. Largent, Jr., who was born during show week. Two older brothers were Tom Largent and W. J. (Willie Joe) Largent, all three of these family members also being active personalities in the Hereford field for many years.

Roy, by then all of six years old, was back at the Fort Worth show in 1908, and so steadfast was he in his attendance at this premier Hereford event of the Southwest during the ensuing decades that his 50th consecutive year at the show was marked by the "Roy L" Hereford show, named in his honor. It was entirely fitting that this 1958 Fort Worth show was one of the greatest in the series, with 42 exhibitors from a dozen states competing–states as far distant

*Residences and ranch buildings of C. M. Largent & Sons at Merkel, Texas, where Roy Largent's career as a Hereford breeder began in a herd established by the father in 1900. Roy was born in the large house in the center of picture. In addition to Roy, the "sons" in the firm at one time or another included Tom, W. J., and C. M., Jr. A lineup of Largent productions of the 1930's is in the foreground—all daughters of Publican Domino.—Smith photo.*

as California, Arizona and Ohio. Thus was the stature of this man, Roy Largent, among his peers in the American Hereford industry.

Roy Largent's father became an exhibitor at the early-day Fort Worth shows when it was known as the National Feeders' and Breeders' Show and won his first purple ribbon there in the 1911 event. Additional purples were captured in the next three Fort Worth shows, to establish firmly the Largent winning tradition. And when the Southwestern Exposition and Fat Stock Show was initiated at Fort Worth in 1918, the Largents were the winners of both blue and purple ribbons.

While he was still a boy, Roy assumed a man's responsibilities at the family ranch. Since the older brothers frequently were on the road with the show herd, he took over the management and feeding of the cattle at home, and his efforts there often were reflected by the performance of the cattle in the showring. This effort made it necessary for him to drop out of school at times, the only exception, he once pointed out, being his senior year in high school. But his father would hire a tutor during the late spring and summer to enable him to catch up with his school work and thus go on with his classmates at the beginning of the next term.

The C. M. Largent & Sons Herefords commanded national attention for the first time when the herd's representatives competed at the American Royal show in Kansas City in 1917 and performed in sensational fashion. In this 1917 event Largent entries won first in the senior yearling bull lineup, second and third on junior yearlings, first in the junior bull calf class, third and fourth on two-year-old heifers, first and junior championship on the senior yearling heifer Shadeland Jewel 2d, second on junior yearling heifer and fourth in the senior heifer calf class, these winnings coming in classes ranging in size up to 26 head. Additionally, the Largents took second on aged herd, first on young herd, and first and fourth on get of sire.

The Largent record in this show was a revelation to northern exhibitors who theretofore were wont to smile at the suggestion that Texas-bred and Texas-fitted cattle could compete successfully with the northern herds. These cattle were bred in the Largent herd, and from then on the Largent name was one to reckon with in the Hereford industry.

Roy Largent once related that his first recollection in connection with the showing of cattle at Fort Worth was when the family show string included a big steer that was declared grand champion of the show. He didn't remember the year, but did recall that the steer was a four-year-old and weighed 1,800 pounds. Quite a task for a boy to feed and fit such a beast, he once reflected, but the result was one that gave him considerable satisfaction.

The greatest showring thrill that ever came his way, Roy Largent went on, was in 1927 when the Largent herd provided the champion bull, Prince Domino 3d, and the champion female, Clo Ann, for that year's Fort Worth show. Many years afterward he remembered these two individuals with admiration and their extraordinary showring performance with a great sense of pride.

Busy as he was with his ranching duties during his growing-up years, Roy Largent still had time for a well-filled extra-curricular schedule. He was a lover of athletics and played football, baseball and basketball, and was a member of the track team in high school. Perhaps the competitive spirit cultivated in these activities was reflected in the Hereford field throughout his career.

After high school, he attended Texas Christian University in Fort Worth, and while doing regular academic work there he also studied voice. His extraordinary talent in that direction was apparent in the fact that he was offered a very flattering proposition to go to Italy for further study. When asked why he had turned the offer down, he replied simply that he "could not even consider the idea of being associated with anything but Hereford cattle." Not many of his Hereford friends were aware that he might well have become a professional vocalist of substantial stature had he so chosen. But it was the

*Roy Largent was gifted with a fine voice and in his student days was a member of the Texas Christian University Glee Club, in which he was a soloist. Roy stood fourth from the left in the second row of the group as it prepared to go on tour.*

*Mr. and Mrs. Roy Largent and their son, David, with a Largent winner at a Sweetwater, Texas, Area Hereford Show. David and his sons, Steve and Mark, now comprise the Hereford firm of Largent & Sons of Wilsall, Montana. Roy and David Largent comprise one of only three father-son combinations to serve as presidents of the American Hereford Association.—Texas Hereford Association photo.*

good fortune of the Hereford industry that he made the choice he did.

As C. M. Largent, Sr., lessened his activity in the operation of the herd, Tom, the eldest son, took charge of the breeding herd and the show string for several years, with Willie Joe as his assistant. Then Tom decided to engage in another line of endeavor and Willie Joe assumed the role of active manager, with Roy being the assistant for several years. C. M. Largent, Jr., moved into the assistant's position when Roy decided to go into business for himself, which he did in 1929 when he moved to Brown County, Texas.

There he became a member of the firm of Largent & Stevens, at Brownwood, Texas, where his partner was D. L. Stevens. He resided in Brown County for 12 years before moving back to his place north of Merkel. He proved to be an effective leader there in advancing the interests of the breed, becoming the first president of the Brown County Hereford Association. The Largent & Stevens operation was represented for several years by animals which were exhibited successfully at southwestern events, primarily, and during this period also the firm's annual auctions of LS Herefords became a dependable source of supply for good, practical cattle.

Mainly the Largent & Stevens herd was based upon the influence of Publican Domino, of intense Prince Domino ancestry, which also was a keystone sire for many years in the W. J. Largent operations at Merkel. Real Prince Domino breeding also figured prominently as Roy Largent strove steadily to produce better Herefords. His aim, he once stated, was to raise cattle with ancestry unsurpassed as to individuality, show records and bloodlines, along with producing ability both on the range and on feed. "Our goal," he went on, "is the production of Herefords that will put on more meat, more evenly, and yield a higher dressing percent of meat where the high-priced cuts are obtained."

His sincerity, ability and energies contributed to his leadership talent which did not go unnoticed among his contemporaries, many of whom were far removed from his local and regional areas. He went forward from his Brown County activity to become president of the West Texas Hereford Association in 1944. State-wide recognition came with his election for three terms as president of the Texas Hereford Association. At the latter group's 1958 meeting during the "Roy L" show, he pointed out that he attended his first Texas Hereford Association banquet during the Fat Stock Show in 1919, and that he had missed neither the banquet nor the organization's business meeting since 1921.

Such faithful and responsible service naturally had brought him wide recognition years earlier, a result of which had been his election as president of the American Hereford Association in 1951. He also served the latter body as a member of its board of directors for six years.

At various times he was a member of the board of directors of the Southwestern Exposition and Fat Stock Show, the West Texas Fair at Abilene, the Sand Hills Hereford Show at Odessa, Texas, and was an honorary vice-president of the State Fair of Texas at Dallas.

The last several years of Roy Largent's activity in the Hereford field were spent as the senior member of the firm of Roy R. Largent & Sons at Merkel,

*Roy Largent, as president of the American Hereford Association, looked on at the right as Jess C. Andrew, second from left, presented a Hereford statuette to Dan Casement when the latter was honored by the AHA at its 1952 annual meeting in Kansas City. Andrew, of West Point, Indiana, was president of the International Live Stock Exposition and a long-time Casement friend. Jack Turner, then secretary of the Hereford Association, is at the left.—AHA photo.*

where his partners were his sons, Rust and David. They kept the Largent name before the public with the show strings which campaigned extensively at events as far distant as Baltimore and San Francisco.

It is appropriate also to state that the Largent name bids fair to continue as a factor in the Hereford field for years to come, even though Roy Largent reached the end of the line in 1968. Rust and his mother, Mrs. Hazel Largent, carried on with the family's LS Hereford operation at Fort Davis, in the Texas Big Bend region, while David broke with the Largent Texas tradition and moved to Montana where his herd became both widely prominent and highly successful.

Further, by the latter 1970's the advertisements of Largent & Sons of Wilsall, Montana, were carrying the names not only of David but also of his sons, Steve and Mark–a new generation of Largents. Of additional significance is the fact that Largent & Sons were co-exhibitors at the 1979 National Western Stock Show in Denver of that great event's champion bull. One more point: David Largent was elected vice-president of the American Hereford Association in November, 1979, and in 1980 followed in his father's footsteps when he was named president–one of the very few father-and-son combinations to serve in that great Hereford leadership post.

Of Roy Largent, it was once written by the late Frank Reeves, livestock editor of the Fort Worth *Star-Telegram*, that he had done some unusual things in a most unpretentious manner, adding:

"He has been anxious to succeed but also he has been anxious for others to succeed. He has given generously of his knowledge and experience so that others could profit from what he has learned. He has been a friend to many."

Reeves' concluding thought with reference to Roy Largent was that "it is not just what you do but the way you do it that counts." The way Roy Largent did it resulted in a great contribution not just to the breed's progress but also in the example he set.

# J. S. Bridwell

WICHITA FALLS, TEXAS

1885-1966

THIS is the story of a man who at the age of 21 left the Missouri farm home where he was born, who then went to Chickasha, Oklahoma, where he worked for a year or two in a grocery store, and then drifted on down to Texas. His intention was to go to Stamford but the train which was to take him there was late, and an impatient Joe Bridwell decided on the spur of the moment to go to Wichita Falls, since the train to that city was ready to go.

This decision was a good one from the standpoint both of the man and that of his adopted home town. And from the standpoint of the Hereford industry, too.

"To say that it was a bonanza for this community when Mr. Bridwell decided to make his home here would be understating the case," the leading newspaper in Wichita Falls pointed out on the day in 1966 when he died. "For not only did this energetic citizen become a major civic leader, heading dozens of groups for community betterment, but also giving generously of his time and money, and making signal contributions to agriculture. . . ."

The newspaper continued: "Wherever Mr. Bridwell applied his boundless energy–in land, in oil, in cattle, in Methodism, in education, in the cause of soil conservation, in responsible citizenship–there arose in his footsteps a monument to his labors. His abilities could be reckoned only in the scope of

*Sale day at Bridwell Hereford Ranch was one of the big days of the Hereford year. Here, Joe Bridwell was speaking at the beginning of his December, 1963, auction, in which 244 head were sold at an average price of $2,125.–Photo by George E. Smiley.*

diversity of his interests–widespread, sweeping and substantial."

When a man of the stature of J. S. Bridwell, as indicated by the above lines, decided at the age of 47 to enter the registered Hereford business, it was a great day for the breed. After briefly engaging in the real estate and insurance business in Wichita Falls, he recognized the potentialities for oil development which was just beginning to gain momentum in this north Texas area. One success in this field led to another, his friendliness and tenacity enabling him to keep a step ahead in land- and oil-leasing trends, and he eventually became one of the most extensive operators in the area.

The next logical step was the formation of the Bridwell Oil Company which started operations in 1927 and grew to encompass some 700 wells in operation in six states, entitling it to rank as one of the largest independent oil-producing companies in the Southwest. As this activity prospered, Joe Bridwell's thoughts by 1930 turned back to the soil, and he shortly thereafter began the enterprise which was to bring him international renown as a Hereford breeder. His land holdings eventually grew to consist of a half-dozen ranches in Texas, two in Kansas, three Mississippi plantations, and one ranch in Colorado, aggregating some 225,000 acres.

He first became actively engaged in the cattle business in an indirect manner. In 1930 he bought 45,000 bushels of oats, and decided that the surest route to making a profit from them was through feeding them to cattle. Accordingly he bought commercial Hereford calves at 5 cents a pound, figuring around $20 a head, from two herds, those of R. V. Colbert & Son of Stamford, and Bryant Edwards of Henrietta, Texas.

He liked the way in which those Hereford calves utilized the oats, and thus cemented forever was his appreciation of Herefords as beef-makers. Two years later the Colbert registered Hereford herd, which was considered one of the breed's leaders in the Southwest, was placed on the market. Bridwell decided to go to Stamford to inspect the herd, taking with him Bryant Edwards, whose judgment of cattle he valued highly. Both men liked the Colbert cattle, all infused with some of the richest of Anxiety 4th bloodlines, and a deal was closed whereby 155 cows, many with calves at side, 50 yearling heifers and six herd bulls were transferred to J. S. Bridwell. He then bought the property at Windthorst, Texas, some 30 miles south of Wichita Falls, which became famous as Bridwell Hereford Ranch. This was so the herd could be maintained closer to his home and he would thus be able to see his Herefords more often.

On that foundation, Joe Bridwell proceeded with a program which shortly attained high prominence. By 1937 a Bridwell carload of bulls won the grand championship of that division at Fort Worth's Southwestern Exposition and Fat Stock Show. Much of the Bridwell herd's show activity through the years was centered around carload-lot competition, since a major accent in the Bridwell merchandising program was to supply the demand for high-class range bulls in quantity. In that direction, the herd's achievements were impressive. Operators of top-flight commercial herds returned again and again, Bryant Edwards of Henrietta, for example, purchasing some 300 head over a period of years.

The first carload of Bridwell bulls to be exhibited at Denver's National Western, conceded to be the greatest carload bull show of all, went west in 1955. That entry won the grand champion award, and every year thereafter in which the ranch was represented the Bridwell entries won either the grand championship or reserve grand championship.

But Bridwell entries for many years also were successful competitors in the individual breeding classes at such leading shows as those in Denver and Kansas City, awards as high as the championship having been captured. Likewise, the Bridwell auction sales held at the ranch never failed to arouse excitement as bidders from far and near assembled. High tide was reached in 1956 when 63 head, all home bred, commanded an average of $4,008. It set a new high mark for the great Hereford state of Texas, and the highest average

*This scene at Bridwell Hereford Ranch, which was located south of Wichita Falls, Texas, near Windthorst, affords an example of the caliber of the cows which comprised the herd, and the sort of youngsters which they produced.—Texas Hereford Association photo.*

attained for the breed anywhere in 1956. No sale of the breed, at least to that date, had averaged as much with as modest a top as the $16,600 peak recorded in that event.

Largely responsible for these successes were two Real Silver Domino herd bulls, "the 203d" and "the 181st," scions of a famous family which traced to the breed-building Prince Domino through Real Prince Domino, the latter generally considered to have been one of the two or three greatest of his sire's sons. These two bulls and their sons and grandsons were instrumental in rebuilding the herd after it, in common with many other top herds in America, became afflicted with a recessive genetic factor known as dwarfism. Joe Bridwell faced the latter problem squarely and discarded at heavy cost to himself all of the cattle of the offending lines. It was characteristic of him that he was undismayed by the disaster which might have lastingly discouraged a man with less vision and determination. And fittingly enough, it was the Real Silver Dominos which led Bridwell Herefords to the highest levels of acclaim in the herd's history. Which in itself is a testimonial to the virtue of courage, for without it Bridwell Hereford Ranch would not have emerged with a reputation as one of America's great Hereford-breeding establishments of all time.

With all of his successes, J. S. Bridwell remained one of the most modest of men. Nevertheless, he found it almost impossible to remain in the background because his leadership qualities and energy were recognized as prime assets in every organization of which he became a member. Within five or six years after his entry into the registered Hereford field, he had been elected a director of the American Hereford Association, and was named its president in 1939. After a brief interim period, he again was elected president of the organization in 1942, and was returned to the same high post in 1943 for yet another term, thus becoming one of the very few men to serve for three terms at the American Hereford Association's helm.

His guidance and advice continued to be sought long after he had relinquished his formal posts, and probably never once did he fail to respond when his counsel was sought. This man, whose searching eyes and rapid gait were almost personal trademarks, possessed a remarkable sense of responsibility, which made him inevitably a leader of men.

He served in various high positions in numerous state and regional organizations of which he was a member. Again and again his services were sought for leadership in his home community. He had many philanthropies, most of which were shielded from public view by his own adamant preference. He built several houses for the Methodist orphanage at Waco, Texas. He financed the construction of the library for the Perkins School of

*Neatness and order always prevailed as the big crowds arrived early for the Bridwell sales to inspect the cattle and make a study of the pedigrees of the individuals which most impressed the various prospective bidders.—Texas Hereford Association photo.*

Theology at Southern Methodist University in Dallas, Texas, and then bought for it a rare and valuable collection of books which he knew the librarian coveted. On his 80th birthday, friends announced a gift of $80,000 as an endowment for the J. S. Bridwell Library of Religion, a non-denominational educational facility at Houston, Texas, operated in connection with the Houston Medical Center.

The Fort Worth *Star-Telegram*, in the 1960's, in a feature article, referred to Joe Bridwell not only as a highly successful ranchman and Hereford breeder, but also hailed him as "Wichita Falls' Number One Citizen," describing him as ". . . a city builder, a master civic organizer, philanthropist, humanitarian, public benefactor and Methodist layman."

How fortunate for the Hereford breed to have had such a man as Joseph Sterling Bridwell as one of its greatest leaders for 34 years! His example offers ample evidence of how much is possible in the Hereford field or almost every other for those who bring vision, vitality, determination and energy to the task.

# J. P. CALLIHAM

CONWAY, TEXAS
1899-1978

METHODICAL and deliberate in his approach to herd and breed improvement, and without the slightest trace of ostentation in his operations, J. P. Calliham accomplished goal after goal in the development of one of the country's leading Hereford herds during its half-century of activity. A native of the Texas Panhandle, who lived near the small town of Conway, not far south of the county seat of Panhandle, Texas, for all of his 78 years, he strived steadily in his own quiet way toward his own personal goals in herd-building, and despite the fact that his progress was widely recognized in the industry his sense of modesty never permitted him to feel entirely satisfied with his herd's accomplishments. Perhaps this provided the impetus for its steady movement forward.

The Calliham family in the Conway area dates back to 1888 when the move westward was made from Collin County, near Dallas. Paul Calliham's father, H. L. Calliham, built a considerable reputation at the family's new location as a producer of mules which were used extensively in the area's farming and ranching activities before the advent of tractors. Up to 50 head a year were raised and sold during the decade before and after World War I, and "J. P.," or "Paul," as he was usually hailed, not only worked with them in the fields in plowing, combining, or doing whatever heavy field work needed attention, but also recalled later that his father sold many mules on credit, taking payment sometimes in grain and fodder. It was partly to utilize this grain that the father was led also to engage in the cattle business.

It was this latter field which aroused Paul Calliham's interest to the extent that in 1927, six years after he was wed, he decided to become a full-fledged registered Hereford breeder. Never mind that finances were scarce in this post-war depression period when he sold wheat for as little as 26 cents a bushel and hogs at $2.55 per cwt. This young man's reputation for integrity already was such that he could borrow from the bank the entire amount needed to purchase 19 heifers and a bull. A mortgage was given on all of his horse-drawn farm machinery.

This set of heifers came from the herd of J. M. Alexander of Claude, Texas, and cost the newcomer $100 a head. His first herd bull was Don Randolph by Beau Randolph 2d, the latter bred in the famous Missouri herd of Gudgell & Simpson. Every cow in the Calliham herd at the climax of his career traced back to this original foundation—a testimonial in itself as to how well those initial purchases were chosen.

The primary inheritance of the entire group was derived from the Gudgell & Simpson source, since the heifers were descended from Beau Chief, a Gudgell & Simpson-bred son of Bright Stanway. The latter was the last principal herd bull of Gudgell & Simpson, and was the top-seller at $3,600 when the herd was dispersed in 1916. Beau Chief was taken to the Panhandle by Jowell & Jowell of Hereford, Texas, and later was used at Claude, Texas, by R. D. Doak & Sons and B. H. Conner.

This first Calliham herd bull possessed a similarly impressive inheritance. His paternal grandsire was Beau Randolph, sired by Gudgell & Simpson's famous Beau President and out of a dam by Dandy Rex by the equally famed Lamplighter. Jowell & Jowell took Beau Randolph to the Panhandle as a four-year-old at a price of $5,000, said to have been the highest price Gudgell & Simpson ever received for an animal. A powerful specimen, possessing both size and quality, he was reported to have weighed 2,700 pounds when he was grand champion bull at the Panhandle State Fair at Amarillo, Texas, in 1916. Shortly after that fair he was sold to Mrs. H. M Pegues & Sons of Odessa, Texas, for $7,000, and proceeded for them to become senior and grand champion bull at the State Fair of Texas at Dallas.

Beau Randolph 2d, from which that first Calliham herd-header sprang, not only was a son of Beau Randolph but his dam was by Beau President by Beau Brummel, these two ranking among the greatest of the Gudgell & Simpson history-makers. In fact, his pedigree is replete with the names of animals which contributed notably to the Missouri firm's reputation. As a point of interest, Beau Randolph 2d, through another son, became the grandsire of Beau Gwen, founder of a Gwen family which performed successfully for many years in Texas and elsewhere, with one member of the family, Beau Gwen 50th, becoming grand champion bull at the Southwestern Exposition and Fat Stock Show at Fort Worth in 1933.

So it is apparent that the Calliham foundation was solidly laid. Later infusions, similarly derived from Gudgell & Simpson ancestry, came through a descendant of Superior Mischief by Beau Mischief, the noted herd bull of Mousel Bros. of Cambridge, Nebraska; Mischief Domino, a grandson of the breed-building Prince Domino, of Fulscher & Kepler fame at Holyoke, Colorado; two sons of The Lamplighter, a straight-Anxiety 4th sire used at Tierra Blanca Farms, Canyon, Texas, following his highly successful service for Mousel Bros.; Sensation Mischief, intensely bred to Mousels' noted Prince Domino Mischief; and Blanchard Return 30th, a grandson of Prince Domino Return, the herd-building sire which brought fame to Dr. C. H. Harris and his Harrisdale Farms at Fort Worth, Texas.

Later Calliham herd bulls, and notable contributors to the herd's accomplishments as it approached its maturity, were C. Real Domino and C. Husker Domino. These two bulls, calved in the 1960's, left a great lot of females in the Calliham breeding herd, which became key contributors to the production of "The Cowmen's Kind," as publicized by Calliham advertisements which were directed to the attention of both the purebred industry and volume buyers of bulls for top-rank commercial herd service.

Since both of these bulls sired so many of the matrons which influenced the Calliham breeding program in subsequent years, it is of interest to note that both represented some of the breed's most productive lines of their era. C. Real Domino, bred by W. J. Dees of Vineyard, Texas, and calved the property of J. P. Calliham, was double-bred to Bear Creek Prince, a

*J. P. Calliham's accomplishments as a breed improver earned many honors for him. Here he displayed a plaque signifying that in January, 1976, he had been made an honorary life member of the Texas Hereford Association. Mrs. Calliham stands at his side. The presentation was made by Roy Herrmann, Caldwell, Texas, then president of the association.—Texas Hereford Association photo.*

*A highlight stop on the tours of the Panhandle Hereford Breeders Association always was that at the Calliham ranch. Here in 1962 a crowd assembled around the family home where a segment of the herd awaited visitors' inspection.—Texas Hereford Association photo.*

noteworthy son of the great Real Domino 51st. The latter, bred in Nebraska by Kimberling Bros., was a grandson of Real Prince Domino and was of Prince Domino descent on both sides of his pedigree. He was the progenitor of the noted Silver Domino line developed at Silver Creek Farms, Fort Worth, Texas, under the management of Jack Turner, which carried on in Turner's own Silver Crest operations after the dispersal of the original herd.

C. Husker Domino was a bull of similarly noteworthy credentials. Bred by Milligan Bros. of Streetman, Texas, he was a grandson of Husker Mischief 1076, scion of a famous line which descended from Husker Mischief, the Mousel-bred bull which contributed so greatly to the fame of the Tequesquite Ranch herd of T. E. Mitchell & Son of Albert, New Mexico. Other strong infusions into the inheritance of C. Husker Domino came from the Colorado Domino line, established on a Prince Domino-Beau Blanchard foundation, by Banning-Lewis Ranches of Colorado Springs, Colorado, and from The Prince Domino, the breed-building sire at the Harrisdale Farms of Dr. C. H. Harris of Fort Worth, to which the grandsire of C. Husker Domino's dam was double-bred.

The hallmark of a genuinely constructive cattle breeder is that through a progression of herd sires, whether home-bred or selected from outside sources, a course of improvement generation by generation is charted. Extraordinary success attended J. P. Calliham's efforts in this direction, as a study of the herd's pedigrees discloses. His selections were made only after great deliberation. His master stroke came in the fall of 1972 when he

*This scene typifies the Panhandle-Plains region in which the Calliham Hereford operations were conducted. A group of brood matrons was assembled here at the water tanks, with a massive herd bull the focus of attention in the midst of the picture.—George Williams photo.*

purchased from Alex Born & Sons of Follett, Texas, the herd-sire prospect which became the great Tex Adv. Onward 3d.

As with his major predecessors in the Calliham herd, this sturdy individual not only became a bull possessing scale and substance in combination with quality, but his inheritance again represented an amalgamation of leading bloodlines and illustrious individuals. His sire, Tex Golden Advance, also bred by the Borns, was sired by their long-time herd bull, HDR Real Onward 166th, a descendant of the Real Prince Domino line through HH Real Domino 203d, a highly successful sire of show and breeding cattle for Hull-Dobbs Ranches of Fort Worth, Texas. Tex Adv. Onward 3d's dam likewise was of Real Prince Domino descent, through her grandsire, Real Silver Domino 203d, a herd bull of renown for Bridwell Hereford Ranch of Wichita Falls, Texas.

While representatives of the Calliham herd seldom were exhibited at other than regional events, the get of Tex Adv. Onward 3d, as the result of the sale of a breeding interest in him to Indian Mound Farm of New Harmony, Indiana, gained international fame through their winnings in the Indiana herd at major shows across the country. So extensive they were that by the middle 1970's this 2,270-pound Calliham product had gained status as America's Number 1 Super Register of Merit Sire. Small wonder thus that sons carried on in the Calliham herd, and that many of his daughters were retained as future brood

matrons, in fact dominating the 125-head cow herd by the latter 1970's.

Although almost 77 years of age, and already a registered Hereford breeder for almost a half-century, Paul Calliham was continuing to look ahead when he bought his final herd bull, Adv. 2005 Domino 421st, at the $5,500 top bull price of the 1976 sale of Indian Mound Farm. Sired by C L1 Domino 2005, a Super Register of Merit sire and one of the most noted of the Line 1 family of Cooper-Holden fame, the dam of this winter calf which had been undefeated in class in five Register of Merit shows and which had climaxed his showring career by winning the grand championship of Fort Worth's Southwestern Exposition was a granddaughter of Tex Adv. Onward 3d and HDR Real Onward 166th. Thus exemplified again was J. P. Calliham's practice of sustaining a constructive breeding program through successive generations.

By the time of Mr. Calliham's passing two years later, only two weeks before his 79th birthday, ranch visitors already were hailing the get of the new bull as "the best group ever seen on the ranch." Happily, a substantial segment of the herd, including the bull from Indian Mound Farm, subsequently became the property of Mr. and Mrs. Don Vance, son-in-law and daughter of Mr. and Mrs. Calliham, who continue to pursue on the Calliham ranch the program so long maintained there.

It was but natural that a man of such attainments as J. P. Calliham in the breeding of Herefords should also be called upon to assume leadership roles in his industry and community. He joined the Panhandle Hereford Breeders Association in 1929 and played a prominent role in it throughout his career. This organization's 1965 sale catalogue featured his picture on its cover page and on that occasion he was hailed as "one of the most distinguished, respected and beloved members" of the association.

At that time he was recognized as having sold not only the first but also the highest-priced bull ever sold in the Gouldy Sales Arena at Amarillo. The popularity of the Calliham Herefords, it was pointed out on this occasion, was "not only due to the quality of the individual animals but also to the reputation for honesty and integrity of the man who produced them."

It is a matter of interest that at this 1965 event four of the eight purple-ribbon winners were Calliham entries, and that one of them set a new price record for the Panhandle sale of $6,700. The previous record for this event was $6,200, established by a previous Calliham bull in 1959. It was said that during a period of more than a quarter of a century, Calliham productions "topped the Panhandle Hereford sale over 75 percent of the time."

Although he sold his Herefords at points as far distant as Denver and Fort Worth at times, and found customers in 29 states and four other countries, his main interest was in supplying a superior product to Hereford breeders and ranchers in the great southwestern ranching region in which he was located. The Top o' Texas Hereford Breeders Association, at Pampa, in the upper Panhandle, thus drew his support from its very beginning. He consigned bulls to its first sale in 1953 and served it as president in 1953. In 1979, a year after his passing, this association's sale was dedicated to his memory.

He also served as a director of the Texas Hereford Association, which recognized him as an honorary life member at its banquet and meeting at Fort Worth in 1976. Likewise, he was a long-standing member and supporter of the American Hereford Association and its programs, and of various progressive youth activities involving Herefords.

It was greatly to the Hereford industry's benefit that a man of his caliber gave so liberally of his time and talent to the breed and its interests for most of a productive lifetime.

# ROY J. TURNER

SULPHUR, OKLAHOMA

1894-1973

No newcomer to the Hereford ranks ever was catapulted more quickly into the limelight than Roy J. Turner of Oklahoma. Although during the preceding year or two, he and the man who was his partner at the outset, Forrest E. Harper, had made herd foundation purchases of some 200 to 250 head of registered Herefords from such nationally prominent herds as those of Foster Farms of Rexford, Kansas; Wyoming Hereford Ranch of Cheyenne, Wyoming; Robert H. Hazlett of El Dorado, Kansas; Taussig Bros. of Parshall, Colorado; Fred Grimes of Kremmling, Colorado; E. F. Swinney of Independence, Missouri, and Walter J. Bones of Parker, South Dakota, it was their determined activity in buying the very cream of the Robert H. Hazlett herd in June, 1937, some six months after the death of the famous Kansas breeder, that boosted the Oklahomans immediately into the Hereford industry's front rank.

In this great Hazlett dispersion sale, one of the landmark events in breed history, and with the skilled guidance of Dean W. L. Blizzard of the Animal Husbandry Department of Oklahoma A. & M. College, the Turner and Harper purchase of 56 lots not only comprised more than twice as many as the selections of the next heaviest buyer in the Hazlett sale, but more importantly consisted of a heavy percentage of the Kansas herd's extreme tops.

*During the 1940's history-making Hereford auctions became the rule at Turner Ranch. In this picture, Roy J. Turner, owner, was pictured welcoming the assembly at the opening of the January 8, 1945, event.—AHA photo.*

These included not only the top-selling herd bull at $6,800, the highest figure that had been paid for a Hereford bull since 1923, but also the 10-head show string which was sold as a unit to the Oklahomans for $18,800, a truly sensational mark in a year when all of the Herefords sold at auction in the U. S. A. averaged only $193. These 10 included the great young show bull, Hazford Rupert 81st, which proceeded four months later to put his new owners' names in the headlines again when he was acclaimed grand champion bull at Kansas City's American Royal Livestock Show.

To back up a bit, Roy Turner was born on the farm in Lincoln County, Oklahoma, which his parents had homesteaded. The early death of his father cut short his farm career at the outset, but after finishing high school at Kingfisher, Oklahoma, he went on to business school and then became a bookkeeper for the Morris Packing Company at Oklahoma City where his work frequently took him into the stockyards as well as behind a desk where he figured the cost of putting carcasses into the cooler. In this job he learned well what it took to turn out good beef, little realizing at the time what this experience might mean to him in later years.

He then worked as a salesman for the Goodyear Tire and Rubber Company and served in World War I. Next came activity in the oil and real estate field in Oklahoma City until he entered into a highly successful partnership with Forrest Harper in 1928 in establishing the Harper-Turner Oil Company which prospered as an independent in the oil-producing business. Neither of these men had forgotten his farm background, and now with money enough to buy land they began to purchase the acreage in south-central Oklahoma which comprised the nucleus of the 10,000-acre tract which was to become the home of one of the world's most noted Hereford-breeding establishments. This was in the Arbuckle limestone region, near the town of Sulphur, which was to gain world-wide fame as "Hereford Heaven."

Roy Turner became the sole owner of the ranch and herd when Harper sold his interest to his partner in the fall of 1938. However, the two continued their 50-50 partnership in other lines, including the oil operation.

It did not take long for Roy Turner personally as well as through his cattle to make his presence felt in the Hereford industry. Named in 1937 to fill the unexpired term of the late Robert Hazlett as a member of the directorate of the American Hereford Association, he was elected in his own right in 1938, and, further, was named to serve as president of the organization for the ensuing year.

An Oklahoma political writer once said of this quiet, unassuming, clear-thinking man, who was destined to become governor of his state in 1947, that his outstanding trait was calm deliberation. That observer added that after taking time to think matters through, Roy Turner was the kind of man who would stand by his decisions. So thoroughly did his fellow Hereford breeders agree with this assessment that he was drafted to serve again as president of the American Hereford Association in 1944 and again in 1945. Only three men in the 100 years of Hereford Association history from 1881 through 1980 served three terms at the organization's helm, a tribute in itself.

Capitalizing from the beginning upon the Hazlett inheritance, Turner Ranch not only won purple and blue ribbons in leading Hereford shows from coast to coast, but also found a nation-wide market for its output. In the very first Turner Ranch auction sale, in 1940, the average price of $1,023 surpassed

*The modern and commodious sale pavilion at the Turner Ranch became something of a Mecca as Hereford enthusiasts from coast to coast were drawn to it for the annual auctions. Roy Turner sat in the front row in the auction box, at extreme right, as three bulls were paraded at the beginning of the 1959 sale.—Denman photo.*

that of any other individual breeder's sale on an offering of comparable size since 1921. Thereafter, the averages went up and up until a high mark of $6,120 was set on 50 head in the 1954 sale. In all of these sales all of the cattle sold were bred on the ranch. So popular was the herd's output that breeding stock through the years was sent out to cattlemen in 43 states and five foreign countries.

Key element in this array of Turner Ranch successes during the first 10 years following his acquisition in the Hazlett dispersion was the show bull, Hazford Rupert 81st. He gained wide renown as the first "million-dollar" bull in America on the basis of sale prices commanded by his sons and daughters, plus the value of his get retained in the Turner breeding herd. Thus was his prepotency appraised by other breeders who sought to infuse into their herds his own type, size and substance, plus his inclination toward early maturity.

If the acquisition of Hazford Rupert 81st in the purchase of the dominant portion of the Hazlett herd was Roy Turner's first 10-strike, the second came with his addition in 1948 of the bull which was to gain international renown as TR Zato Heir. Bred by the Patterson Land Company of Bismarck, North Dakota, and tracing in the top line of his pedigree to Zato Rupert, a half-brother to Hazford Rupert 81st, both having been sired by Hazlett's Hazford Rupert 25th, he also carried a strong infusion of Prince Domino inheritance, particularly through Advance Domino and Prince Domino Heir.

Nicking marvelously well with the predecessor lines in the Turner herd, his appearance came when the Hereford industry was passing through a traumatic period induced by the genetic plague of dwarfism which afflicted some family lines. TR Zato Heir proved to be the right bull at the right time as he proceeded to infuse, through his sons and daughters, a vital element of stability to the Hereford industry because of the proved soundness of his genetic inheritance.

This alone was sufficient to produce an enormous demand for his get, which was reflected in the high-averaging Turner Ranch sales during the 1950's. In addition, the Zato Heirs also proved their ability to follow in the footsteps of earlier Turner showring winners, crossing wonderfully well with the Hazlett families used in the Turner breeding program previously.

Probably no exact record of the winnings to Turner Ranch in major fairs and expositions, as well as Register of Merit contests, exists, but that hundreds of ribbons and scores of silver trophies were captured there is no doubt. In top-flight Register of Merit competition alone Turner Ranch accounted for more than two dozen purple-ribbon awards on bulls it exhibited, and perhaps twice as many more were won on bulls exhibited by other breeders but sired by bulls bred at Turner Ranch. Likewise in get-of-sire winnings, one of the most prized of all awards in breeding cattle competition, Turner Ranch entries were victorious at least 15 times during this period in Register of Merit contests. But perhaps of even greater significance is the fact that during this same span, blue-ribbon get-of-sire entries sired by bulls of Turner Ranch ancestry were shown by other exhibitors on at least 51 occasions. It was this kind of breeding performance that enhanced the Turner reputation among Hereford producers across America.

It should be cited that Roy Turner was an enthusiastic advocate of youth programs. He inaugurated a series of field days at the ranch for Oklahoma 4-H club members and Future Farmers of America. He provided the cattle for use in judging contests, and also personally donated the funds for cash prizes and trophies for the various winners. He considered this activity a fine investment in the future not alone of livestock agriculture but also in America itself.

As the years passed, Roy Turner in the spring of 1963, some 10 years before his death, announced a complete dispersion of his herd, then consisting of more than 1,600 head. A massive catalogue of what was hailed as "The History Book Dispersion" was issued. But a few weeks before the announced date, the entire layout–ranch and Herefords–was sold to Winthrop Rockefeller, then of Morrilton, Arkansas. It undoubtedly was the largest catalogue ever put out for a sale that was never held.

But the Turner legacy lives on, particularly in the minds of those who witnessed the quarter-century during which Roy Turner became a genuine benefactor of the Hereford breed and the Hereford industry. His contribution was both timely and substantial.

*Roy Turner was very generous in his support of youth activities, the Turner Ranch junior field days being annual highlight events over a period of several years. Five judging rings were operating simultaneously on this occasion as more than 450 FFA and 4-H club members tested their judging skills on four classes of Turner Ranch bulls and one of heifers.–AHA photo.*

# James M. McClelland

SULPHUR, OKLAHOMA
1902-

It has been the Hereford breed's good fortune throughout its history in America that men of ability and dedication have ably served it as herdsmen and managers. Some observers have commented, in fact, that these men applied themselves more selflessly than if they had been the actual owners of the animals involved. That many hundreds of them have been genuine benefactors of the breed there can be no doubt.

One such man, who stands as a symbol of others of like talent and interest, has been James M. McClelland of Sulphur, Oklahoma. Virtually his entire lifetime has been spent in working either as herdsman or manager with outstanding herds of beef cattle, mostly Herefords. His first activity with registered Herefords dates back to January, 1923.

The fact that Jim McClelland's home town was Maple Hill, Kansas, may have had something to do with his inclination toward Whitefaces. Two of the noted commercial Hereford operations of his boyhood days there were the ranches and feedlots of W. J. Tod and H. G. Adams. They were more or less constantly "in the news," and whether or not young Jim was consciously influenced or not may never be known with certainty, but it is well within the realm of reason that those early impressions may have made more of an imprint than he could possibly have imagined.

In any case, he had reached the age of 21 when he took that 1923 job and began to learn the tasks involved in working with Herefords. He remembers well that it was at a farm named Terrace Lake Hereford Park, straight south of downtown Kansas City, Missouri, near the village of Hickman Mills. Something like a quarter of a century later this area was absorbed by an expanding Kansas City, but in the '20's it was the home of a prominent Anxiety 4th-bred herd owned by Wallace Good, Jr., of Kansas City and Harry H. Rogers, a businessman of Tulsa, Oklahoma.

Jim was an ambitious young fellow and when an opportunity to become herdsman for James Mulvihill & Son at Cumming, Iowa, presented itself he accepted, and began new duties there in January, 1925. He fitted a small show string for his new employers, which was exhibited at some nearby county fairs, and then at the Iowa State Fair in Des Moines, and at the 1925 American Royal in Kansas City.

With this bit of experience under his belt, he then went back to Terrace Lake as full-time herdsman. The next year, 1926, the Terrace Lake show herd campaigned extensively during the fall, beginning at the Missouri State Fair in August and continuing through the State Fair of Texas at Dallas in October. At the latter event, a deal was struck whereby the entire Terrace Lake herd was sold to R. V. Colbert & Son of Stamford, Texas, and Jim McClelland went along with the herd to become the Colbert herdsman. There he fitted the Colbert show herd, and also started the preparation of a carload of bulls for showing.

He also fitted the cattle for a Colbert auction held in 1927, at which a considerable number of the show cattle were bought by John M. Gist of Odessa, Texas. This West Texan arranged with the Colberts for McClelland to go along with the cattle and take charge of the Gist show string. Gist exhibited at several Texas events during the fall of that year, winding up at the Southwestern Exposition and Fat Stock Show in Fort Worth the following March where the young herdsman led out the Gist heifer which captured the event's junior championship of the female division. In that same show the Colberts exhibited the grand champion carload of bulls, which had been started on their way by the young Kansan before he left Stamford.

McClelland's Texas years ended later in 1928 when drouth conditions in West Texas led many herd owners there to ship their cattle to the Kansas Flint Hills, where better pasture conditions prevailed. So he came back home with a trainload of Texas cattle, plus a lot of experience which was to stand him in good stead throughout his future years.

That same summer he was engaged by Tomson Bros. of Wakarusa, Kansas, to handle their Shorthorn show herd at the Kansas fairs and at the

American Royal. Following the Kansas City event, he spent a long year as herdsman with the Wheatland Farms' Angus herd of James Hollinger of Chapman, Kansas. And it was, indeed, a long year, as McClelland recalls that the herd made 16 shows. But it was a very successful season, with top winners including the grand champion steer at Denver's National Western at Denver in 1929, and a female that was champion or reserve champion at 12 of the 16 shows in which the herd competed.

When that season's shows ended in March, 1930, Jim marked time until June, but with his energy and experience it seemed certain that the wait would not be a lengthy one, even though a national depression was just then getting well under way. But he knew he now had a wealth of practical experience that would be invaluable when the right opportunity presented itself.

That opportunity came when R. J. Kinzer, then secretary of the American Hereford Cattle Breeders' Association, called him from Kansas City and asked that he be at Kinzer's office the following morning. McClelland was there at the appointed hour and waiting in the secretary's office to talk with him was E. D. (Doc) Mustoe, manager of Foster Farms at Rexford, Kansas, an extensive Hereford-producing and wheat-raising property owned by Benjamin B. Foster, a Kansas City lumber executive. A strong foundation had been laid for the Foster Farms herd and the program outlined during the course of the interview was of such promise that McClelland accepted the challenge to become herdsman.

Started in 1919, the herd was predominantly of straight Anxiety 4th ancestry, and already was rated among the country's leaders, with the breeding of Foster Anxiety, a grandson of Bright Stanway, and Promino by Advance Mischief in the forefront. McClelland proceeded to do with extraordinary success the job for which he had been employed, as Foster Farms exhibited dozens of purple-ribbon winners at leading fairs and expositions during the succeeding eight years.

One of these was Beau Stanway 2d, which was adjudged the Denver show's champion bull in 1935 and then brought the Denver sale's top price that year in going to the Matador Land and Cattle Company which long had operated extensively in West Texas. Beau Beauty, a grandson of Promino, won heavily, including numerous championships, between 1934 and 1937, as did his half-sister, Mabel's Beauty 35th, female champion in 1937 at Denver and in 1938 at Fort Worth.

These were just three of many individual Foster winners fed and handled by McClelland, but he once indicated that he was proudest of all of the group winnings of the Foster Herefords. These included such victories as the American Royal's "best 10 head" in 1937 and the first-prize get-of-sire group at the 1938 Chicago International, among numerous others. Jim McClelland by now had amply proved his mettle not only in fitting and showing top-notch Herefords but also for his ability in picking the top prospects from the successive calf crops. He was ready for the next step.

It came at the International in Chicago during the 1938 show. Again it was Secretary Kinzer who was the catalyst in bringing him together with Roy J. Turner of Oklahoma City and Sulphur, Oklahoma, who had just bought the interest of his original partner, Forrest Harper, in the Harper & Turner ranch, which thereafter was to gain international renown as Turner Ranch. "After the question-and-answer part of the interview was over," McClelland once related, "I was asked if I thought I could fill the job as manager of the ranch and herd, and would I accept? My reply to both parts of the question was 'yes'."

During his tenure at the helm, extending from December, 1938, to July, 1960, Turner Ranch compiled a record of show winnings in top competition which may never be surpassed. In Register of Merit competition alone, Turner Ranch won a total of 55 championships, 46 reserve championships, 517 first-prizes, 355 seconds, 268 thirds, 229 fourths and 180 fifths. Jim McClelland, in reflecting back to those years, recalls that the herd also was shown extensively at the state fairs, and expressed confidence that if those events were included the above figures could easily be doubled.

Turner Ranch's annual auction sales became widely heralded criterion events during this era. The highest average recorded was in 1954, when 50 head averaged

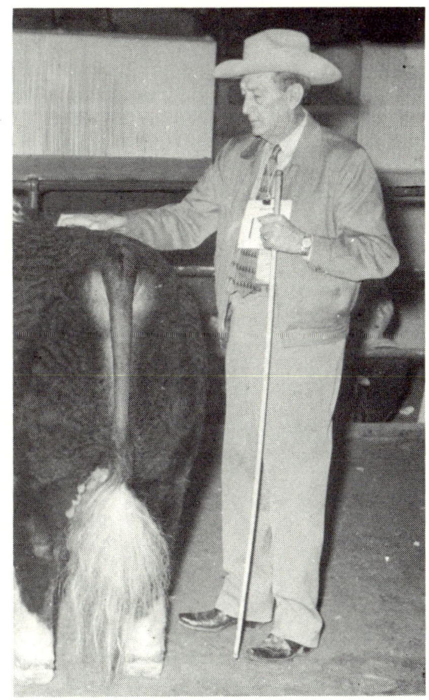

*J. M. (Jim) McClelland was meticulous in his presentation of the show animals which were in his charge in the herds he managed. Here, show stick in hand, he was working with TR Zato Model 61st which became the 1960 Denver show's champion bull. Turner Ranch, which he managed for a long period, and Healey Bros. of Davis, Oklahoma, were joint owners of the bull when the award was won.*—Hereford Journal *photo.*

$6,120, with a top price of $45,100, paid for a one-half interest in TR Royal Zato 27th, reserve champion bull at the previous fall's American Royal, thereby establishing a valuation of $90,200.

This bull was a grandson of TR Zato Heir, and Jim McClelland may be pardoned if he takes greatest pride of all in having, in essence, "discovered" this bull in the show herd of the Patterson Land Company of Bismarck, North Dakota. TR Zato Heir, then named Zato Heir L. 22d, appeared as a calf at the Chicago International of 1947, and made such a great impression that McClelland later journeyed to North Dakota to study him and his ancestors in greater detail. The fact that the youngster carried a fair amount of Hazlett breeding was significant because the Turner herd had been largely founded on Hazlett bloodlines. But beyond this, there was a spark of greatness in the calf himself that intrigued McClelland.

He telephoned his employer, Roy Turner, who by then was governor of Oklahoma, asking that he come to North Dakota to see the calf. Turner demurred because of the press of his official duties. Although McClelland's mind was made up, he insisted that Turner should make the trip, because the infusion of this calf's influence would mark a distinct turn in the Turner Ranch program. So Turner went to Bismarck and the deal was consummated in short order. The price: $15,000. The rest became Hereford history as the bull, renamed TR Zato Heir, became a paramount influence on the Hereford breed in subsequent bovine generations.

TR Zato Heir's discovery and purchase came at a propitious time. He fortunately possessed the genetic potential to infuse not only quality into the herd but also his service at Turner Ranch resulted in a renewed sense of stability in the Hereford industry which then was facing the dwarfism problem, which the TR Zato Heirs helped greatly in solving. By his discovery of a bull calf, and discerning in him something which he believed the herd and the breed could use with utmost benefit, Jim McClelland earned a lasting place in Hereford annals.

When the Turner operations were curtailed in 1960, McClelland resigned and became consultant for Healey Bros. Flying L. Ranch at Davis, Oklahoma, for a few months. In January, 1961, he accepted an offer by Ben Smith to become general manager of his Lost River Ranch at Klamath Falls, Oregon. A very successful six years were spent there, during which numerous champions produced and fitted under McClelland's direction were exhibited at the major Hereford events of the far-western region.

In January, 1967, he resigned the Lost River post and he and his wife, Ruth, returned to Sulphur where they had purchased a home in which to retire. But Jim couldn't stay retired. Four months later he resumed his association with

*Few men in breed history judged as many important Hereford shows during a busy career as did Jim McClelland. This picture shows him relaxing between classes as onlookers, as well as the judge, await the arrival in the arena of the next class. Jim McClelland was equally proficient in making selections of breeding cattle in the herds he managed, and in planning the matings in them.*—Hereford Journal *photo.*

the program of the ambitious Healey brothers, and continued with them until 1971. He recalls having shown for the Healeys their champion bull at the 1968 Fort Worth show, and remembered that this was exactly 40 years after the Colbert carload of bulls had been Fort Worth grand champions.

But even then he couldn't stay retired. During the four or five years immediately prior to 1980 he assisted James R. Jack of Sulphur in establishing a Hereford herd, after which Jim related that Mr. Jack is "well on his way to producing cattle that will meet today's requirements, with some further advances on the plus-side."

The "Jim McClelland Register of Merit Hereford Show," held in 1974 in connection with the Tulsa State Fair at Tulsa, Oklahoma, honored him for his lifetime of contributions to the Hereford industry, and also recalled to the minds of many who attended this event that during his career he not only had exhibited winners at most of the country's great shows, but also that he had officiated as judge at such national events as those at Denver, San Francisco, Chicago and Portland, plus many leading state fairs.

But in addition to his showring activities, both as exhibitor and as a judge, Jim McClelland looks back with equal satisfaction to the results of his management duties including the selection of breeding stock, decisions on matings, selections of herd sires and the like. This is, as he pointed out, where it all begins. Unless the "makings" are there, he asserts, the feeding, fitting and showing skills avail little or nothing. Thus, his all-around record stands as a reflection of what many herdsmen and managers have contributed in varying degrees to the advancement of the Hereford breed in America.

# ARTHUR D. WEBER

MANHATTAN, KANSAS

1898-

MANY educators in the fields of animal husbandry and animal science have contributed notably to the progress of the Hereford breed through the years, but none more so than Dr. Arthur D. Weber, who spent most of his active lifetime on the staff of the institution at Manhattan, Kansas, which in due course came to be known as Kansas State University. A man of sober judgment based on a combination of common sense, education and practicality, his guidance, advice and counsel benefitted the Hereford industry again and again during the middle to latter years of the 20th century.

Life began for Weber, who came to be familiarly known, because of his precocious maturity, as "Dad," even in high school days, on his father's farm near Muscotah, Kansas. His keen interest in livestock matters led him to the then Kansas State Agricultural College at Manhattan, from which he was graduated in animal husbandry in 1922. He earned his entire way through college by working in the beef-cattle barn for three-and-a-half years and by his employment for eight months at Miller Hereford Ranch at nearby Belvue, Kansas.

For a considerable portion of that time, he shared room and board with another student, J. J. (Jerry) Moxley, who was destined later to become a Kansas Hereford breeder of substantial stature and considerable reputation, at Council Grove, Kansas. They were on the team which in 1922 won first in the collegiate livestock judging contest at Denver's National Western Stock Show and thus claimed for the college permanent possession of a $500 challenge trophy. A few weeks earlier Weber had been second-highest individual judge in the contest at the International Live Stock Exposition at Chicago.

His education and the experience gained in handling the college cattle led, following his graduation, to his employment first as working manager of Cameston Farm near Lenexa, Kansas, where the owner, Charles M. Keith, a Kansas City businessman, maintained herds of purebred Herefords and Poland China hogs. After some two years, he returned in 1924 to Manhattan to become an instructor in animal husbandry. He responded in 1926 to a call from the University of Nebraska at Lincoln and spent five years there in research and teaching before returning to Kansas State to re-establish a connection that was to continue through the remainder of his active years.

Thus it was that in 1931 he became professor in charge of beef cattle research and teaching at Kansas State, continuing in that position until 1944 when he was named head of the Department of Animal Husbandry. Time out had been taken in the late 1930's for him to earn his Master of Science degree and his Ph.D. from Purdue University at Lafayette, Indiana. He continued at the helm of Kansas State's Department of Animal Husbandry until 1950, then proceeded to become Dean of Agriculture, in charge of all research, teaching and extension in agriculture, and Director of the Kansas Agricultural Experiment Station, continuing until 1960. During the latter portion of this period, he served from time to time as Acting President of the University, and shunned opportunities to assume the presidency of other similarly prestigious institutions.

Dr. Weber's stature in his field was recognized by the executives of the Ford Foundation when he was selected in 1959 as a member of the team of U.S. experts designated to advise the government of India on how best to meet the food crisis which confronted that nation's teeming millions. Thereafter he returned again and again to India, having been granted a leave of absence by Kansas State University to serve as head of the Ford Foundation's project which was designed to increase India's food production.

He relinquished that position on July 1, 1962, to return to Kansas State University as vice-president for general administration and director of all of the University's programs with India as well as with Justus Liebig University, the University of Munich, and the People-to-People program. He also played a significant role in counseling international students. During his last three years as a staff member at the Manhattan institution, Dr. Weber was

*No judge of beef cattle of his era had greater prestige among his contemporaries than Dr. A. D. Weber of Manhattan, Kansas, second from left, who spent most of a fruitful lifetime at the institution which came to be known as Kansas State University. He judged all of the fed steer classes at Chicago's International Live Stock Exposition, then the greatest event of its kind, for 13 consecutive years, a period unapproached by anyone else in the event's history. At left in the picture was Albert K. Mitchell, Hereford breeder of Albert, New Mexico, then serving as chairman of the board of directors of the International, while at their grand champion steer's halter, holding the championship trophy and ribbon, were Pete and Sue Secondino of West Terre Haute, Indiana.—Abernathy photo.*

chief-of-party for Kansas State University's faculty in India and special advisor at Andhra Pradesh Agricultural University there, which encompassed six formerly independent colleges of agriculture, veterinary medicine and home science. It is greatly to Dr. Weber's credit that as a result of the programs which he headed, India became more nearly food-sufficient than ever before in its modern history.

All of the above is indicative of the stature of the man whose wisdom benefitted the Hereford breed again and again during a span of 40 to 50 years. His competence was proved scores, if not hundreds, of times by his talks given at breed conferences, field days and the like, as well as on an individual basis in assisting breeders with their problems. But equally valuable was his common-sense, unflappable approach to problems, which usually led to their solution. As early as the 1930's he was contributing articles of a basic nature to *The American Hereford Journal.*

Always one to engage in adventures of the mind in projecting practical possibilities built on scientific research, Dr. Weber correctly foresaw potential benefits which could result from progeny and performance testing in beef cattle, based upon procedures first proposed in 1932. Well before comprehensive investigations on a broad scale were far advanced, Dr. Weber recognized the value of this work to the industry and undertook the responsibility of reducing to a level of general understanding the complexities involved.

By the early to middle 1950's, he was freely expressing his deep conviction that in the years ahead progeny and performance testing would have "a terrific impact on the beef cattle business. I don't know," he went on, "when economic survival will force the majority of beef cattle breeders into such testing, but it is to be hoped that cattlemen will not wait until economic survival forces them to do so."

Dr. Weber correctly anticipated that as awareness grew among commercial cow-herd owners of the economic benefits of improved gaining ability and feed efficiency possible through adherence to testing programs, an ever increasing demand for purebred bulls with performance records would develop. He added that it would be desirable for the associations to assume responsibility for technical direction and supervision of the programs for the respective breeds. The American Hereford Association, fortunately, heeded his admonition.

A Weber article in *The American Hereford Journal* in 1937 dealt with the subject of heredity and touched upon a phenomenon which later beset the Hereford and other breeds when he referred to the abnormality of so-called bulldog calves, commented on their genetic basis, and discussed the means by which such monstrosities could be eliminated. Thus, his counsel was called into action as the Hereford industry successfully battled the plague of dwarfism during the mid-years of the 20th century.

Dr. Weber attained international stature as a judge of all beef cattle breeds. After a beginning in the 1930's with state and regional fairs, he officiated time and again at most of the Hereford Register of Merit and other national Hereford events, beginning with the Denver show in 1944.

He kept no record of the shows which he judged, but in 1979 he related that such a list would include most of the major events throughout the United States, plus some in Canada. "My experiences in judging at the National Western Stock Show in Denver," he stated, "were quite diverse, including individual breeding classes, carlot bulls and carlot feeder cattle."

He went on to add that a highlight of his Hereford judging experiences occurred in Argentina when he officiated in 1949 in placing the breeding classes at the world-renowned show at Palermo Park in Buenos Aires. Dr. Weber was reported to have been the first American ever invited to serve as a judge there.

A year earlier, in 1948, Dr. Weber judged all of the individual fed steer classes at the Chicago International, thus becoming the first American judge ever to officiate at this then most prestigious steer show in America, if not the entire world. During the preceding 48 Internationals, the steer division had been judged by men from England, Scotland, Canada, Argentina and Ireland. So ably did Dr. Weber perform his task that he continued to judge the steer division of the International for 13 consecutive years, a span unapproached by anyone else.

Dr. Weber became in 1953 the first American ever invited to judge the supreme beef cattle championships and make the award of the Duke of Norfolk Cup at the famous Smithfield Show, held annually at Earls Court, London, England, an event which was inaugurated in 1798. When this show was over, Dr. Weber, one of the most modest and unassuming of men, came in for an unusual bit of personal glory, as indicated by the Duke of Norfolk in remarks presented to the assembly following the judging.

"I can assure you," he said, "that it was a joy to watch the manner in which, with supreme confidence and knowledge, Dr. Weber set about his job. Having heard the remarks around the ring, I doubt whether any other Smithfield show has ever ended with such complete and universal praise concerning the man who carried out the judging."

The Duke of Norfolk's conclusion was confirmed when Dr. Weber was invited to return three years later to again judge the cattle championship classes at Smithfield. In announcing his selection, the English agricultural magazine, *Farmer and Stock-Breeder*, stated: "It will be remembered that Dean Weber acted in a similar capacity at the 1953 show, when he gave one of the best displays of judging seen since the turn of the century."

Many honors were bestowed upon Dr. Weber by many organizations affiliated with the livestock industry during his illustrious career, but the most prestigious perhaps came in 1952 when his portrait was unveiled during the International show in the Saddle and Sirloin Club's gallery of history-making leaders in animal husbandry and livestock improvement. Hundreds of stockmen and educators filled the banquet room to listen as a distinguished array of speakers paid tribute to him for his many years of scientific contributions to the livestock industry. Indeed, as was pointed out by a spokesman for the sponsoring American Society of Animal Production, of which he had once been president, the award was given not for the facet of his life by which the public knew Dean Weber best–livestock judging. Rather, it was given for his scientific contributions in the field of livestock improvement. In pointing to his background of solid research, it was emphasized that he had been the author of more than 60 publications which discussed advances in the field of animal science.

Another first for Dr. Weber came in 1960 when the American Royal's beef cattle show was dedicated to the Kansas professor. Never before had such distinction been given. It was cited that the honor was in public recognition of his contributions not only to the Hereford industry but also because he had been a life-long benefactor of the beef-cattle business in general.

It is impossible to cite all of the honors that came his way, but a living memorial which undoubtedly was as greatly appreciated as it was richly deserved came when Weber Hall, a key component of the facilities of the Department of Animal Science at Kansas State University, was named in his honor. In reviewing his achievements at a banquet when he took leave of the University in 1968, that institution's then president, James A. McCain, lauded him when the Weber Hall Library was dedicated on the campus to which he came as a student in 1917 and left in 1968 as vice-president emeritus.

In case Dr. McCain did not publicly recognize Dr. Weber's contribution to the Hereford industry on that occasion, the induction of this practical-minded man of science into the Hereford Honor Gallery served well this purpose. The Hereford breed is a better breed, a more practical breed, and a sounder breed as a result of his constructive influence.

*At its 1975 unveiling at Kansas State University, Dr. Weber stood beside the bust of himself in Weber Hall, headquarters of the institution's Department of Animal Science. The plaque below it reads: "Arthur D. 'Dad' Weber, stockman, teacher, scientist, administrator, philosopher. A recognized international leader in agriculture. A man of great integrity. Presented by stockmen and friends. March 6, 1975."*

# JOHN J. VANIER

SALINA, KANSAS
1897-1980

TIME has long since proved John J. Vanier to have been a visionary of the first rank. This trait, combined with a keen business sense and an industrious nature, enabled him to rise from meager beginnings on a farm in Pawnee County, Nebraska, to the ownership of many properties, including the largest registered Hereford herd in the United States, at CK Ranch at Brookville, Kansas, west of Salina, in the central part of the state. In fact, the CK in the ranch name stands for Central Kansas.

But it must surely have been a vision of what might be that kept the young John Vanier firmly on course as he rose step by step from his Nebraska boyhood. He did not finish high school but did take a four-month course in stenography which enabled him to go to work as a boy for the A. B. Stevens Hide Company in Kansas City for $7 a week. He did his work so well that three years later, in 1917 when he was 20 years old, he accepted a $60-a-month job with the E. D. Fisher Commission Company in the Board of Trade building in Kansas City. Here he came in daily contact with the milling industry and envisioned that this should be the direction of his future. A year later he went to work for the Abilene Flour Mills at Abilene, Kansas.

World War I intervened a few months later and John Vanier enlisted in the Marine Corps, his stenographic experience leading to his assignment in a recruiting office. But when the war ended he got back on track quickly by returning to the mill in Abilene where he managed to save $90 in his first year from his salary of $75 a month.

By later standards a very small sum, this $90 was sufficient to provide John Vanier's essential start. With it he bought an option on 30 shares of stock, the local banker being sufficiently impressed by the young man's no-nonsense, business-like nature to loan him the money to pay for the stock. Vanier had accumulated 70 shares of stock by 1922, and when an opportunity came for him to sell out for $21,000 he grasped it. Because he already had looked ahead and knew what he wanted to do with the money.

He used it as a down payment in the purchase of a controlling interest in the Western Star Milling Company at Salina, Kansas, some 25 miles west of Abilene. Its management had grown lethargic, he thought, and he believed that an energetic infusion of new plans and procedures might bring it out of the doldrums. His vision of that potential came true as the Western Star became the centerpiece of the subsequent extensive Vanier operations.

It was all upward after that, the modest and unassuming John Vanier, who much preferred to remain in the background, once conceded. And, he felt there was no secret about the key to his business success. "There is no question in my mind but it's work," he once told a visitor, adding, "and I don't mean the hours between punching in and punching out." He summed it all up in these words: "Work and save and buy and sell."

When land prices in the early 1930's reached a level just about the lowest in modern history, John Vanier foresaw that the purchase of ranch and farm property should be a good investment. This conviction led to his purchase in 1933 of 5,600 acres of prime ranch land some four miles west of Brookville, a one-time trail town in the days of the great Texas cattle drives. Well grassed hills and valleys and an unfailing source of water that gushed from rocky hillsides made this an almost ideal ranch property, as earlier owners had already discovered.

One of these was Ed Root who, from the Terra Cotta stockyards, on the property, shipped as many as 10,000 southwestern grass-fat steers a year to Kansas City and other markets during the approximately 30 years of his ownership, a period when he was widely recognized as a leader in the industry. This region is the Kansas counterpart of expansive ranching areas in Wyoming or Montana, and CK Ranch under Vanier ownership fits right into the picture as an entirely practical cattle-breeding establishment rather than a show place.

By 1936 John Vanier had laid the foundation for his CK Herefords. Breeding stock came from such herds as those of Otto Fulscher of Holyoke,

*In this picture, Mr. and Mrs. John Vanier, left, were receiving from George Lazear, then manager of Wyoming Hereford Ranch, the "Bob Lazear Memorial Trophy" symbolic of the carlot bull show's grand championship won by their CK Ranch entry at the 1962 National Western Stock Show.–AHA photo.*

Colorado; Fred C. DeBerard of Kremmling, Colorado; Wyoming Hereford Ranch of Cheyenne, Wyoming; W. C. Windsor of Boonville, Missouri; Roy Fahlstrom of Concordia, Kansas; S. S. Spangler of Hutchinson, Kansas; the George Godfrey Moore estate of Topeka, Kansas, and Fred Grimes of Kremmling, Colorado, among others.

But the master stroke in the CK herd's beginnings was the acquisition of the entire Fred Grimes herd in 1937. One of the cows in that Grimes purchase was carrying the calf which was recorded by CK Ranch as CK Onward Domino. Double-bred to Prince Domino through Onward Domino and Dandy Domino 2d, he became the first history-maker in the Kansas herd as the progenitor of the CK Challengers, the CK Cruisers and the CK Crustys.

Rapid expansion of the breeding herd came during the early 1940's until, by 1946, the ranch's fall sale catalogue listed 33 bulls in the herd-sire battery. A majority of these were home-bred but others represented several leading Prince Domino lines from other sources.

This continued to be the Vanier pattern until 1952 when TR Zato Heir 40th, a son of Turner Ranch's noted TR Zato Heir, was bought in that year's National Western sale at Denver for $35,000, incidentally the same sale in which CK Ranch sold CK Crusty 46th, reserve champion of the 1952 Denver show, for the same sale's top price of $41,500. For CK, the new herd bull established a noteworthy line CK Ckatos which figured significantly in the herd's output for several years.

By this time, CK Ranch had long since proved that its entries could hold their own, and then some, in the breed's hottest competition. Champion bulls bred in the herd had captured Denver championships as early as 1948 and 1950, with dozens of other high awards coming in other leading shows.

All the while the expansion program of the breeding herd continued until annual registration statistics provided evidence that the CK herd had become one of the breed's largest, as well as one of the best. By the early to mid-1950's upwards of 1,000 calves were being registered annually, with the peak figure being 1,523 head.

This quality-quantity combination made CK Ranch a popular source of supply for owners of high-class commercial herds, and many operators of leading ranches all the way from Florida to Texas to the Rocky Mountain West made annual treks to buy in the CK auction sales. For many years, dating back to around 1940, semi-annual auctions were held, but in more recent times an annual fall sale involving substantial numbers was the ranch's primary auction outlet for bulls in quantity. That these bulls did what John Vanier anticipated they would do was amply indicated by the successes scored by the get of these bulls in carlot feeder cattle competition, and then in the feedlots in which they and their mates were finished for market.

CK's owner then tested the mettle of CK bulls themselves in the greatest of all carlot bull shows, that at Denver's National Western, with the result that within a span of 14 years, beginning in 1951, CK bulls won the grand championship award eight times. And in some of the years when CK did not win, the supreme purple was captured by exhibitors whose entries were sired by CK-bred bulls.

It was during this period that John Vanier again demonstrated his vision with a prediction that the market price for fed beeves eventually—and within a relatively brief span of time—would reach 50 cents a pound. With prices at the time of his comment standing at only a little more than half that amount, there were many who scoffed. In fact, few did not. But with confidence in his own judgment, he proceeded with his program which eventually led to Vanier purebred operations not only at Brookville, but also at Hunter, Dorrance, Herington and Manhattan, all in Kansas, and commercials herds at Alcova, Wyoming; Trinidad, Colorado; Keyes, Oklahoma, and Ellis, Kansas. Cows involved in the five registered Hereford operations rose to some 2,000 head.

Thus it was that when the fed cattle market soared to levels not only of 50 cents a pound, but far above that figure, CK Ranch, through Vanier's foresight, was ready and waiting.

*The size and caliber of CK Ranch's registered Hereford operation was indicated by this picture of a portion of the herd-bull battery. Stretching beyond the bulls was an expanse of the CK range in central Kansas.—Smith photo.*

This was as it had been in the milling business which at the height of his operations included not only two flour mills at Salina, but also other milling plants, terminal elevators, alfalfa dehydrators, soybean plants, warehouses, a major macaroni and spaghetti processing plant, a livestock feed manufacturing plant, and the like. In connection with the latter, he conceived the idea of giving 4-H club and FFA members an opportunity to buy Hereford calves, pigs and, occasionally, other items, such as sewing machines for the girls, using as "money" in their bidding at this annual auction event coupons inserted into the bags of feed from the Vanier mills. These sales provided many a youngster with a start in feeding or cattle-raising projects, and in all likelihood led many toward careers in the livestock industry. John Vanier continued his activity in the flour and milling field until these enterprises were sold around 1970 to the Archer-Daniels-Midland Company. As to the Herefords, his personal activity lessened during the decade which followed, but his interest continued on and on until his passing early in 1980.

When the specter of dwarfism arose in the Hereford industry in the middle years of the 20th century, John Vanier was one of the first to meet it head on with what he termed a Check and Double-Check program. Based on animal inheritance and owner experience, it was designed to reassure CK's customers as to the genetic soundness of the herd's produce, and was symbolic of John

Vanier's firm approach when confronted by a problem, of whatever kind.

His foresight and leadership qualities were demonstrated in yet another way after his election in 1946 to the board of directors of the American Hereford Association. After serving out his first three-year term, he was reelected in 1949, and it was during this period that a decision was made to erect in Kansas City, on a commanding site overlooking the stockyards area, a spacious and imposing new headquarters building for the Association.

Because of his broad business experience, his judgment and foresight, as well as his dedication to the Hereford industry, John Vanier was chosen as chairman of the four-man building committee which saw this project through to a successful conclusion. The building at 715 Hereford Drive stands as a monument not only to the Hereford industry but also to the board of directors which initiated the project, to the building committee, and to the building committee's chairman, John Vanier.

In all of his years in looking ahead, it is unlikely that anything in a business line could have been more gratifying to John Vanier than the election in 1978 of his son, Jack Vanier, to the board of directors of the Association. For many years Jack Vanier had been in immediate charge of CK's cattle and ranching operations. Thus the Vanier influence in the Hereford field, already a major factor for a generation, carries on.

*This wintry picture was significant in its indication of the hardihood of these CK Ranch Hereford cows and their youngsters in withstanding the cold and snow. It was apparent from this 1965 picture than these were "outdoor" cattle.—Jack Vanier photo.*

# Earl H. Monahan

HYANNIS, NEBRASKA

1899-

EARL MONAHAN was born into the Hereford ranching business, and scarcely a day of his life ever was spent very far from it. His home since he was born has been at the headquarters of the ranch, some 15 miles northeast of Hyannis, Nebraska, far famed in Hereford circles as the Circle Dot. And he would have had it no other way. For both by instinct and inheritance he was always a rancher and cattleman.

He was only three years old when this inclination became boldly apparent. He had overheard his father, James H. Monahan, founder of the Monahan cattle interests, comment casually that he was going to have to get a bunch of steers out of the meadows and take them to the hills for summer grazing.

Young Earl then and there demonstrated at this early age that he was anything but a procrastinator when it came to getting a needed task done. Astride his favorite horse, and without saying a word to anyone, he rode into the meadows, gathered up a rowdy bunch of big steers and put them into the summer pasture.

In due course he went back to the ranch house and casually mentioned the chore he had performed. An earlier writer said that his mother, though a pioneer, nearly fainted when she heard his remark. She needn't have worried. Riding a horse and herding cattle came as naturally to Earl Monahan as for a new-born calf to nurse, or a colt to kick up his heels in springtime. It remained that way with him throughout a busy lifetime.

With his natural proclivities so ably demonstrated at that tender age, it probably came as no surprise to the Monahan neighbors when his father, in 1921, turned over the day-to-day management of the Monahan operations to Earl, then 22 years old, and just home from his studies at Boyles Business College at Omaha which he attended after graduating from Hyannis High School.

The son relished the challenge and during the years which followed guided the tripling of the family ranch holdings located in three counties in the heart of the Sandhills; expanded and improved the commercial Herefords for which the Monahans came to be known far and wide; established a registered Hereford herd for the primary purpose of supplying bulls of the desired caliber for service in the family commercial operation, and, seemingly, never took a backward step.

It has been aptly said of Earl Monahan, a giant in the cattle business at five feet, seven inches in height and weighing in his prime not much more than 130 pounds, that he combined the qualities of a cowboy's cowboy and those of an astute businessman, willing to open his mind to new ideas but aiming to shun the fads that come and go. Although modest and self-effacing by nature, he began early in his ranching career to demonstrate exceptional leadership qualities which could not be hidden from his comtemporaries. When responsibilities thus engendered came to him, he accepted them and left his mark upon most of the organizations which he headed during the course of a busy career.

His achievements at home and away belied a jest once expressed to a visitor who inquired as to Earl's father's philosophy in molding the beginning of one of Nebraska's largest ranches. Earl responded to the query in these words: "He liked to get out early and come in late."

"Have you followed the same philosophy," the visitor asked.

"No," he said, with a twinkle in his eyes, "I sort of liked to get out late and come in early."

This remark is a characteristic example of the sort of dry wit and humor of which his friends are well aware.

A few words on the Monahan beginnings in the Sandhills may be appropriate. Earl's father, James H., then 16 years old, reached the Sandhills when he accompanied his mother there in a move from southwestern Iowa in 1887. The belongings which they brought with them included two old Hereford cows and their calves–which might be said to have been the beginning of the Monahan Hereford holdings. The following spring, young

*Earl Monahan as he appeared in 1938 while on a bull-buying expedition in search of bulls to sire future generations of the famed Circle Dot feeder cattle.—* Hereford Journal *photo.*

James Monahan went to work for his grandfather, the late Robert Christopher Abbott, who had come from Ireland some years earlier. As a hand on the Abbott ranch, his wages were $12 a month. But the wise old grandfather agreed to pay him in cash only enough to buy such clothing as was necessary. "The rest ye'll take in heifer calves," Grandfather Abbott said. And so it was.

James Monahan worked on the Abbott ranch for about nine years, by which time, he once said, he had "quite a little bunch of cattle." Looking ahead, James Monahan in 1894 had homesteaded 160 acres in a nearby broad valley where a sod house was constructed. He was now ready to strike out on his own.

Later a full section was homesteaded, under provisions of the Kincaid Homestead Act. With his savings, he purchased additional acreage whenever he could. Thus was made the start of what first was known as the C-A ranch, and was later to gain fame as the Circle Dot.

Before Earl Monahan established the ranch's purebred Hereford herd, he had become nationally known as a buyer of top-flight registered Hereford bulls for use in producing Circle Dot feeder cattle. In that pursuit, he followed in the footsteps of his father who realized early in his career that such sires were vital in producing the kind of cattle that he envisioned. James Monahan bought his first purebred Hereford bulls at Grand Island, Nebraska, and continued thereafter, until Earl assumed the management role, to add superior sires to the herd.

Although Earl Monahan obviously had made numerous noteworthy bull acquisitions during the 1920's in his endeavor to improve further the herd's feeder cattle output, his first such purchase to be reported in *The American Hereford Journal* seemingly occurred in 1934, when he selected 20 head, ranging in age from a year to a year-and-a-half old, from Fred C. DeBerard of Kremmling, Colorado, for $200 a head. This purchase of bulls for commercial-herd use made news for two reasons: First, the price was remarkable in a year when all of the registered Herefords sold at auction had averaged a mere $114, and secondly, the Colorado seller's reputation as one of the West's leading breeders of registered Herefords was well established and widely known. In 1937, an additional 10 DeBerard bulls were added to the Circle Dot herd at $350 around. Again in 1938 DeBerard bulls were bought. By the latter 1930's the Circle Dot sire battery included more than 70 DeBerard bulls.

The grand champion carload of bulls at Denver's National Western Stock Show in 1941 was purchased from Wyoming Hereford Ranch of Cheyenne, this set of senior calves representing an investment of $600 a head in a year when all of the bulls sold in the Denver yards averaged $239. This and other purchases of superior bulls made from outstanding herds of registered Herefords continued to intensify the focus on Earl Monahan and the Circle Dot commercial Herefords.

Something else happened during the 1930's. The Monahan steers, sired by this sterling set of bulls representing some of the breed's top-producing bloodlines, began to attract national attention at the marketplace. Many carloads of them were fed out by Fred Attebery of Mitchell, Nebraska, and week after week they established weekly top prices on the Chicago market. Some elite restaurants even listed on their menus "Attebery Steaks" as their very best dinners.

The Monahan commercial heifer output also was in demand. Those which the Monahans passed over in making replacement selections for their own breeding herd went to the same Iowa buyers for more than 30 years consecutively. This is conclusive evidence of complete customer satisfaction.

Although never extensive exhibitors, the feeder calf output of the Monahan herd made its appearance on occasion at the Omaha Feeder Calf Show and won more than one ranch's share of the top awards. And more than once, when fed out by others for competition in fat steer shows, they proved that the confidence of their exhibitors had not been misplaced.

Earl Monahan took the step which led to the establishment of the ranch's registered Hereford herd in 1948 when he purchased 24 heifers of Advance Domi-

*Perfectly "at home" aboard a favorite mount, as from the days when he was a very small boy, Earl Monahan was pictured here as he was ready to head out from his ranch home for a look at some of the Circle Dot Herefords.—Monahan photo.*

no breeding, along with a Star Domino bull, from Wyoming Hereford Ranch. Other acquisitions from major sources followed, with a prime step in the herd improvement program being made with the purchase in 1956 of four bull calves from T. E. Mitchell & Son of Albert, New Mexico, and four bulls from Linda Lambert of Mosquero, New Mexico. The "Son" of the former firm, Albert K. Mitchell, was Linda Lambert's brother. One of the Lambert bulls, in particular, became a notable contributor to the progress of the Monahan purebred operation, with one of his sons and many of his grandsons coming in time to dominate the Circle Dot breeding program, and others going out to good purebred and commercial herds throughout the central and northern plains. This Hereford family represented the Mischief line which descended from Mischief 5th, a grandson of Bright Stanway and Beau Mischief which T. E. Mitchell & Son bought from Mousel Bros. in 1921.

A progression of promising herd-bull prospects representing some of the breed's major families followed. During the 1950's these included representatives of the Zato Heirs, Domino Heirs and Regents of Real Prince Domino descent, along with other Real Prince Dominos, among others. Later infusions represented the Regulator and Mill Iron lines, the Dr. Onwards, an Evan Real bull tracing to Evan Mischief, a Dan Mischief, a member of the Sam Donald family and the Super Register of Merit bull, Britisher Prince 20Z, tracing on his sire's sire to the English-bred Wetmore Halflight and from a Britisher-Real Prince Domino dam.

Featured more recently have been bulls of the Line 1 family and a double-bred grandson of Coloradoan F. R. Carpenter's Viscount 4th. Key sire among these has been CL 1 Domino 2005, a Super Register Merit sire, co-owned with Indian Mound Farm of New Harmony, Indiana. As his name indicates, he is a Line 1 bull on his sire's side, while his dam traces to Prince Domino 9th through the Line 10 family, another inbred line developed at the Miles City experiment station.

In a sense, the Monahan operation is unique. Since the purebred herd was established mainly as a source of sires for commercial herd service, the latter is used as a proving ground for the output of the purebred herd. If a new bull does not sire calves good enough for use in the Circle Dot commercial operations, he is not considered adequate as a sire of cattle good enough for buyers of Monahan purebreds and thus is promptly discarded.

When their own range bull needs were supplied, Earl Monahan and his son, Jim Monahan, who had moved into a management position by the late 1950's, decided on the auction route to market their surplus bulls. Their first sale, in 1967, averaged $1,064 on 72 head. Rather than being elated by the excellent average, Earl Monahan was somewhat dismayed. "I'm afraid that the rather high average might tend to discourage many commercial men who think they can't get good bulls at prices they can afford," he said. That comment was indicative of Earl Monahan's broad and perceptive outlook.

It has been the breadth of his view of the livestock industry, and of agriculture in its entirety, that made Earl Monahan such an outstanding leader in local, state and national circles. He was appointed to the board of the American Hereford Association in 1953 to fill out an unexpired term, and then was elected for two more terms, concluding with his election to the AHA's presidency in 1960, when he was hailed as "a natural cattleman and keen businessman." He served as president of the Nebraska Stockgrowers Association, and on various committees of that organization, and in 1955 was honored as the state's outstanding rancher and designated to receive the G. F. Swift Centennial Founders Award. He was a director of the American National Cattlemen's Association, and a member of its National Livestock Tax Committee. He long has been active in the Sandhills Cattle Association.

When the state of Nebraska assumed responsibility for brand inspection of cattle in 1941, Earl Monahan was listed as one of the five original members of the Nebraska Brand Committee, and then was re-named to a second four-year term, serving for a total of eight years. Nor did he shun his local community responsibilities, no matter how busy he may have been.

That the Hereford industry was greatly benefitted by the counsel and guidance of an able man of such accomplishments is beyond all doubt. And also beyond doubt is the statement that the Hereford breed is a better breed because of his life-long interest and activity in it. If possible, he would express a modest disclaimer at this point, which would serve merely to exemplify another Earl Monahan trait.

*A sample of the bull output of the registered herd lined up for this picture a few years ago at headquarters facilities of the Monahan Cattle Company. Note the Circle Dot brand on end of the barn.—Rocky Mountain photo.*

# CHARLES DESCHEEMAEKER

BOYD, MONTANA

1917-

IN the entire history of the Hereford breed in America, only a relative handful of Hereford auction sales have grossed more than a million dollars. The first came in 1951. One of the more recent, as of this writing, was that of Charles Descheemaeker of Boyd, Montana, held in the fall of 1979. Not only did it surpass the magic-million mark, but it did so by a healthy margin of more than an additional third of a million—a conclusive tribute to his constructive effort and extraordinary success as a progressive-minded Hereford breeder.

Charles Descheemaeker was 15 years old when his father, August Descheemaeker, and the latter's partner, Alfons Croes, both natives of Belgium, jointly founded the herd which was moved to the ranch at Boyd in 1944. There the Descheemaeker-Croes operations were continued until 1959. During that year Charles Descheemaeker purchased the herd in its entirety. But Charles had grown up with the herd and had been involved with Herefords throughout his lifetime. Indeed, he had bought three heifers on his own as far back as 1934 when he recalls having paid Stole Herefords of Livingston, Montana, $50 a head for them—considered a "chancy venture" by his concerned family and friends.

His personal involvement with Herefords also was manifested in another activity. This was maintaining the records and the bookkeeping on the Descheemaeker-Croes Herefords. But not until he bought the herd was it enrolled in the Montana Beef-Performance Association, since the herd's original owners had not been greatly concerned about this activity. But Charles bought a set of scales and almost immediately became a leading advocate and practitioner in the budding field of production testing of beef cattle.

His willingness to speak out in behalf of this program at a time when it was largely being ignored by a majority of breeders, and even derided by some, led this writer to invite him to describe and explain his testing program, which he did in the middle 1960's in an article entitled "Today's Breeder Must Do It Better—Production Practices Which Once Were Adequate No Longer Suffice in the Registered Hereford Business." He was probably the first registered Hereford breeder to expound in detail upon his personal experience in production testing in *The American Hereford Journal.*

Excerpts from that pioneering article follow: "Competition from our fellow breeders and the other breeds are creating problems which have to be met if we are to continue as successful breeders. . . . Production testing, along with a general over-all improvement program, requires more careful planning, more record keeping and, naturally, more time spent with the cattle. . . . Production testing has proven to be a step in the right direction as it provides an additional tool for us to use in the improvement of our herd. We continue to select for quality, conformation and the type we can easily sell to today's modern cattlemen, and at the same time test the performance of the animals so we can have more recorded information which we can use when culling, selecting replacements, mating the cows with the sires from which they will raise the best calves, and also to keep tab on how the herd bulls are performing in comparison with each other. . . . A fair test merely requires that the animals in the group to be tested have been raised in nearly the same environment, [and that they be] grouped according to age and the type of pasture they were nursed on."

Additional excerpts follow: "I believe too much emphasis is placed on rate of gain and not enough on the index rating when checking the results of a test. A reasonable, moderate rate of gain is sufficient and the emphasis should be on how the individual animal compares with the average of the group on a daily-gain index and yearling-weight index. This shows percentage-wise how much better or worse he rates than the average of the group he was tested with." Descheemaeker added that brood cows or herd bulls whose calves rated well below the average soon went onto the list of breeding animals to be culled, and their influence thereby eliminated from the herd.

*As an early exponent of performance testing practices, Charles Descheemaeker frequently was called upon to discuss the techniques which he used and the benefits derived from the program.*—Hereford Journal *photo.*

Descheemaeker concluded his presentation in these words: "Our goal is to produce the kind that will wean heavy, thereby having a high nursing gain, continue to gain fast on a minimum 140-day feed test, and consequently have the ability to convert feed to beef efficiently and rapidly every day until they are ready for market. Our expanding population," he continued, "depleting land resources, and competition from other breeds and sources of red meats, will soon require that these traits be bred into our cattle to the highest degree." Another result of the Descheemaeker program was the production of Herefords of greater size and scale than had been the rule in the immediately preceding decades.

But Charles Descheemaeker was well on the way toward being "discovered," especially in his region of the West, before the above explanation of his program appeared. His bull calves were prime attractions in the 1963 auction at the Midland Beef-Cattle Performance Testing Center at Billings, Montana, and the first one he ever sold there to go into a registered herd topped that sale at $3,125 in going to E.P. Reese's Pine Creek Hereford Ranch at Salmon, Idaho. The price was the highest recorded to that date in that event. This youngster at just a little more than a year old weighed 1,165 pounds, and with a recorded daily gain of 3.30 pounds on a 140-day test exemplified to an impressive degree the result of Descheemaeker's program. In the Reese herd, the new bull's performance matched his promise as his sons in three consecutive Pine Creek auctions set the pace, 36 of them averaging $1,544 in 1968 and 21 head reaching a mark of $2,148 in 1969, with a new high point for Pine Creek of $8,000 being recorded in the latter event. The commercial ranchers to whom most of these youngsters went were impressed not only with the calves themselves but also with their massive, 2,500-pound, Descheemaeker-bred sire.

With his reputation well on the way to becoming established in the eyes of the Hereford public, Charles Descheemaeker was on the receiving end of a stroke of good fortune in 1967, although at the time it may not have seemed that way. In that year's Midland Test sale, in April, a bull calf named RC Mischief D. 4th, just a couple of months over a year old, was sold to go to a purchaser from Custer, Montana. The price was moderate—something less than $1,000. But in the middle of the summer a phone call related the rather dismaying information that the buyers did not feel that the young bull was settling the cows satisfactorily. Realizing full well the value of satisfied customers, Descheemaeker promptly refunded the purchase price and brought the bull back home. On the home ranch he proceeded to perform normally and within a few years earned recognition as one of the breed's great all-time history-makers.

The timing of RC Mischief D. 4th's return to the Descheemaeker herd was exactly right. By the time his first calves there began their development, intense attention throughout the industry was beginning to focus upon not only the performance characteristics which these youngsters displayed magnificently but also upon more size and scale in breeding stock, which these calves promised to provide, and which their immediate ancestors possessed. The basis for this was a Hereford Type Conference held at Madison, Wisconsin, in the summer of 1969, which placed the focus as never before upon these elements as essential factors in the Herefords of the future. It was a stroke of good fortune for Charles Descheemaeker who stood ready to meet at least in part the ensuing demand, and for the breed that his foresight enabled him to be in a position to do so.

By the spring of 1972 a string of 31 bull calves by RC Mischief D. 4th were features of the annual Descheemaeker sale and averaged $3,602. The top calf, RC Mischief K. 66th, brought $20,000 from Derry Hereford Ranch of Wood, South Dakota, and M. & M. Hereford Ranch of Hermosa, South Dakota. Two years later the Derrys sold a son of "K. 66th" for $31,000. At about the same time, M. & M. sold a half-interest in a bull calf by the same sire for $35,000.

This recital could go on and on, but the facts cited are sufficient to make the point. RC Mischief D. 4th was the right bull at a most opportune time when he "went to work" for an imaginative and forward-looking owner.

As the major influence in the Descheemaeker breeding program during the herd's final decade, it may be appropriate to summarize RC Mischief D. 4th's record, beginning with the bull himself, a 2,500-pounder at maturity. His statistics in the well-regarded Midland Beef Performance Test were: Nursing Index, 122; Gain Index, 130, and Yearling Index, 112. These figures reflected his superiority in performance traits of vital economic importance.

Through his herd service he became the highest-ranking Super Register of Merit sire as recorded by the American Hereford Association; the highest-ranking Carcass ROM sire for two years; the sire of 13 bulls which became ROM sires; the sire of five Super ROM sires; and on the National Reference Sire Program, he sired an 8-Star Super Sire, a 6-Star performance bull, a

5-Star, a 4-Star and two 3-Star performance bulls. RC Mischief D. 4th also scored several times with the highest-gaining sire groups at the Midland testing station.

That the Descheemaeker output not only possessed the size and scale that the industry sought but also the conformation and quality to compare favorably with that of other leading herds was demonstrated repeatedly during the middle and latter 1970's. Examples abound. A carload of Descheemaeker bull calves, all "D. 4's," in 1974 captured the grand championship award at the greatest of all carlot bull shows, that at Denver's National Western. A son of this history-maker was champion at Denver's great Register of Merit show in 1977.

The line's capacity to breed on and on, a splendid asset in itself, was demonstrated by the Denver performance of "D. 4" descendants. A grandson was reserve champion in 1973 and a great-grandson won the top award in 1978. Two different granddaughters accounted for Denver championships in 1974 and 1977. Among his more prepotent sons was RC Mischief K. 6th, later re-named Winrock D. 4th, which sired many winners at leading shows for the Turner Ranch Division of Winrock Farms of Sulphur, Oklahoma, following his acquisition from Descheemaeker at a price of $14,500 in the Midland Test sale of 1970. One of these was the 1973 Denver reserve champion bull.

Descheemaeker breeding also produced numerous champion individuals and carloads of steers, including the top-winning Hereford individuals at Denver in 1978 and at Fort Worth in 1979. These were raised in the

*Charles Descheemaeker lined up here with a string of junior bull calves which won the grand championship award in the carlot bull show at Denver's National Western in 1974. All were sired by the legendary RC Mischief D. 4th.*—Hereford Journal *photo.*

*A few of the brood matrons in the Descheemaeker herd and a couple of sturdy calves which exemplified their produce. Beyond them the sprinkler irrigation system is in action.—AHA photo.*

commercial herd of the Descheemaekers' son, Larry, on his ranch at Grassrange, Montana.

While RC Mischief D. 4th was the most famous of the lot, 13 other RC bulls featured in an advertisement of the herd's 1979 dispersion sale weighed an average of 2,520 pounds. All were not of identical breeding but they did have a common paternal ancestor, Evan Mischief, as recognized on the letterhead of Descheemaeker Herefords, in this line: "RC Evan Mischiefs." [RC stands for Rock Creek, the name by which the Descheemaeker ranch originally was known.]

Whence came Evan Mischief, and by what route did he become a factor in Montana's Hereford industry? He was bred by H. H. Forney & Son of Lakeside, Nebraska, and was triple-bred to Clayton Domino, a grandson of the noted Kimberling Bros. herd bull, Onward Domino, he a son of Prince Domino. At 18 months of age, Evan Mischief was adjudged champion bull of the 1942 Cornhusker Futurity show at Broken Bow, Nebraska, and then was purchased at the $4,000 top price of the sale by A.C. Bayers, founder of Bayers Hereford Ranch, Twin Bridges, Montana. It is of interest to note that the judge of that show was John C. Burns of Fort Worth, Texas, one of the country's best known cattlemen, then manager of the famous Burnett 6666 Ranches in Texas, and later for several years a consultant in the operation of the equally noted Mill Iron Ranches in the same state.

In the Bayers herd Evan Mischief very shortly established his credentials as a sire of show winners in the state's two major shows, those at Great Falls and Billings. In these events six of his get won 25 ribbons, including two championships, first on get of sire and first on three bulls. In his 1946 sale catalogue, A.C. Bayers made this significant comment about the Evan Mischiefs he had exhibited: "About the only fault the judges found with them was that they were too large for show cattle. We considered this a great compliment."

In his 1947 sale, A.C. Bayers sold the 18-months-old Diamond Mischief 3d by Evan Mischief to Sidney Fraser, Jr. of Reedpoint, Montana. He was by no means one of the sale-toppers, going at $975. Placed in service at Fraser Hereford Ranch his get included FHR Mischief 42d which, in turn, went to the Descheemaeker ranch. There, he became the key foundation sire of the RC Evan Mischiefs, one of his sons, RC Mischief 116th, becoming the grandsire of RC Mischief D. 4th.

Another point of interest from an inheritance standpoint may be the fact that the dam of RC Mischief D. 4th was a granddaughter of Advance Domino E. 3d, tracing to Advance Domino 21st by Advance Domino 13th, and thus a half-brother to the DeBerard-bred Advance Domino 20th and Advance Domino 54th, progenitors of the Line 1 family of Herefords.

There is more that might be written, but this is sufficient to establish Charles Descheemaeker's role in marking a major turning point in the history of the Hereford breed as the three-quarter point in the 20th century approached.

HOLDEN

COOPER

# LESLIE HOLDEN
VALIER, MONTANA
1911-

# JACK COOPER
WILLOW CREEK, MONTANA
1917-

THE story of Leslie Holden and Jack Cooper, two men who dared to be different, is one of the most unusual in the modern history of the Hereford breed. These two half-brothers, the latter a native of Montana and the former born at Reno, Nevada, but a Montana resident since the age of four, bought their first registered Herefords at about the same time–in the late 1940's–with cattle of a type and bloodline not then highly popular in breed circles.

Both had similar goals–to improve the gaining ability and feeding efficiency of the cattle they produced. Both became pioneers in the use of production-testing techniques to advance their programs, becoming charter members of the Montana Beef Performance Association in 1956, at a time when this organization was regarded as controversial. Each man operated his own herd independently but they cooperated even to the extent of aiming to wean their calves on the same day and striving toward about the same

*Intent upon proceedings in the auction ring, Jack Cooper contemplated the animal before him and perhaps considered its potential.*—Drovers Journal *photo.*

weight-gain goals. They also regularly exchanged herd bulls in seeking to avoid any possible risk from breeding too closely. And, until 1979, they joined forces through a 14-year period to hold annual spring auction sales which became history-makers of the first rank.

Because their lives, aspirations and Hereford-breeding careers have been so closely intertwined—although their ranches for more than 25 years have been separated by some 200 miles—Jack Cooper and Leslie Holden were recognized jointly in November, 1980, for induction into the Honor Gallery of the American Hereford Association's Hereford Heritage Hall. So it shall be here.

Both grew up under the tutelage of Frank Cooper, father of Jack and stepfather of Les, in ranch country at Willow Creek, Montana, 50 miles west of Bozeman. This is in the mountain foothills region near where three Montana rivers join to form the mighty Missouri. Both chose ranching as a career although Holden worked for the government for a couple of years until he could get started on his own. Cooper stayed on at Willow Creek, while Holden first rented a ranch at Willow Creek, and then moved to another at Townsend, Montana, which he also rented.

Then, in 1955 he bought a ranch on the treeless plains at Valier, Montana, some 200 miles north of Willow Creek, and only about 50 miles south of the Canadian boundary. When someone once inquired why he had chosen this rather remote location, he frankly replied, "It was the only place where we could make the down payment fit." This indicates the sense of modesty, lack of pretension and easy-going manner that strikingly marks both of these men.

Both men had a keen perception of the direction in which they wanted to go. They were unimpressed by the cattle which were generally winning in the shows during the 1940's and '50's and by the breeding programs which produced many of those cattle. Their sights were set on the commercial bull market. They were most familiar with the ranch trade and considered that the men running such operations would be their principal customers.

Their conclusion was that the range producers were looking for bulls of well above the average size and weight-for-age then prevalent, for soundness of feet and legs which would enable them to travel well, and for inherent

*Leslie Holden gazed toward the far horizon of the range from which his herd's product spread far and wide across the country.*—Drovers Journal *photo.*

capacity to put on pounds more rapidly and efficiently than most cattle of the time. Thus it was that Jack Cooper and Les Holden bypassed the most popular type and bloodlines of the times to start their own individual purebred herds with foundation stock from a source then frequently frowned upon by other purebred herd operators and even by breed officials.

Their choice was the so-called L1, or Line 1, Herefords, and so well did these two men succeed in pursuing their ideal that within a decade or so other seedstock producers began to bid against the Montana commercial ranchers for the output of the herds of Holden and Cooper. And as the word spread purebred breeders from points far beyond Montana's boundaries began to come and look and be impressed.

Line 1 was terminology little known other than to a limited extent in Montana and among livestock scientists when these two men bought their first registered Herefords. And some of those who did recognize the significance of the term shunned the cattle thus identified as products of the work of the "impractical" men in charge at the U.S. Range Livestock Experiment Station at Miles City, Montana, now widely known as the Livestock and Range Research Station.

But results have long since proved that the researchers at Miles City knew what they were doing. Their concern from the inception of their program was not so much with the present as with the future—not so much with the 1940's and '50's as with the 1970's and '80's and beyond. And they could afford to ignore the pressures in behalf of the blockier, tidier type of cattle most generally popular at the time the Miles City program was initiated as they kept their eyes set on their long-term goal.

So it was that Jack Cooper and Les Holden, with their personal preference for cattle of the rangier kind already well established, turned to Miles City as a source of foundation material which they felt certain could hasten the progress of their own herds. Cooper bought his first Line 1 foundation stock when he got 40 females at the Miles City Station in 1947, and in the same year Holden purchased his first Line 1 bull for use in his grade cow herd.

From that point, the latter eased into the purebred field by reducing numbers in his commercial cow herd and replacing them with registered

Herefords. The first of the latter came in his acquisition of two consecutive heifer calf crops from Carl Kieckbush of Townsend, Montana. It was a fortunate circumstance that these heifers mostly were sired by Advance Mixer 60th, which traced back to Advance Domino 13th, the basic progenitor of the Line 1 Herefords at Miles City, and also that the dams of the Kieckbush heifers likewise were of Advance Domino 13th descent. More later about this inheritance.

Buttressing their own personal instincts as Cooper and Holden undertook the development of their breeding programs on a Line 1 basis was the fact that the recognition and intensification of performance traits had been a paramount factor at the Miles City Station almost from the very outset. That Cooper and Holden could begin with cattle already pointed in that direction was a fortunate circumstance—fortunate for the men involved, for the Hereford breed and for the beef-making industry. Fortunate, too, from the viewpoint of all of these elements was the fact that constructive guidance was available, if desired, from an animal scientist at the Miles City Station who had worked with the Line 1 cattle since their second generation.

This was Dr. Ray R. Woodward who was involved with the Line 1 project probably longer than any other person and who possessed an awareness of its potential certainly beyond that of any non-professional. It so happened that Mrs. Woodward is a full-sister of Jack Cooper and a half-sister of Les Holden, and it was entirely logical that those men should have visited from time to time with their brother-in-law concerning their hopes, plans and aspirations. And it was just as logical that they should have consulted with him as to directions they might follow toward the fulfillment of their ambitions.

While obviously interested and gratified in the successes which have attended the personal effort of Cooper and Holden, Dr. Woodward emphasizes that his contribution to their ventures has been relatively minor. He admits to encouraging their interest in investigating the Line 1 Herefords, but then states with finality that "basically the decisions were theirs and they deserve the credit for the success they have enjoyed."

By the time the first Cooper and Holden purchases of Line 1 Herefords were made at Miles City, this line had been closely bred there for approximately 15 years. The program began with the purchase of Advance Domino 20th and Advance Domino 54th, both by Advance Domino 13th by Mousel Bros.' Advance Domino, which the Station had purchased at Denver in 1933 from Fred C. DeBerard of Kremmling, Colorado. Advance Domino 20th had won first prize in the senior bull calf class at that year's National Western Stock Show for DeBerard and in the sale which followed he went to the Miles City Station at $500. No other bull that year sold there for more.

The first daughters of Advance Domino 20th were mated with Advance Domino 54th and "the 54th" daughters were mated with "the 20th." The line thereafter remained closed to outside breeding, with descendants of the two foundation sires continuing their input into the program generation after generation.

Combined with this inbreeding program was a practice of intense selection for important economic traits. The performance testing and severe culling practices pursued at Miles City gave Cooper and Holden insight into the benefits to be derived from the practices followed there, and intensified their decisions to conduct their own operations on a "performance" basis.

"I have always been an advocate of production testing," said Jack Cooper recently, adding: "In selecting a herd sire, I rate his performance record No. 1, and, No. 2, he has to be of good and sound conformation." Both men are in full agreement that performance testing is essential in all herds that are aiming for economic success, and both became charter members of the Montana Beef Performance Association when it was formed in 1956. Both also have been enthusiastic participants in the American Hereford Association's Total Performance Records program from its outset in 1964. And both credit their progress in large measure to the guidance provided by these comprehensive records programs and the selection and culling decisions which they induce.

They are well aware that inbreeding or closebreeding may well make it necessary to cull severely to avoid the fixing of undesirable traits at the same time the desired characteristics are being strengthened. But they add that good growth and good conformation are not necessarily antagonistic elements. "In fact," Holden once wrote, "in many feed trials those individuals making the best gains often also had superior conformation." Adds Cooper: "Our experience has been that if a bull is tops in weight gain, he is tops in conformation." Holden continued in these words: "Performance testing can be applied to any herd, breed or bloodline, and used to identify or select those individuals having superior genetic characteristics for making better growth and gains."

Although bulls of Cooper-Holden ancestry weighing from 2,250 to 2,500 pounds or thereabouts are not uncommon, Holden is not one who overemphasizes mature weight although he does recognize that there is some correlation between mature size and rate of gain. "However," he has said, "one should not lose sight of the fact that the ability to grow to desirable market weight with an acceptable degree of finish in the shortest possible time is the most important economic trait of any beef animal." The end result in a purebred herd or in a commercial herd in which performance-tested bulls are used is "more pounds of better beef from each animal unit," he adds.

Both Holden and Cooper operate cow herds of around 160 head on ranches of some 4,000 acres. They have consistently utilized Line 1 bloodlines from the beginning, getting their first herd bulls at Miles City when herd-sire prospects there could be bought for a few hundred dollars because of the limited demand from other purebred breeders. Since then they have raised herd bulls for their own use, have traded herd sires, and, occasionally, have gone back to Miles City to make purchases.

Cooper cites CH Domino 053 as a bull which he raised as having been an extraordinary sire, with many descendants carrying on not only in the home herd but in others around the country. The most important sire used originally in the Holden herd, according to the owner, was L1 Domino 159, bred at the Miles City Station. He mentions other Line 1's from the same source as having contributed to the breeding program, plus others which have been used as a result of the Holden-Cooper exchanges.

The customary practice of both men is to keep a bull in service no longer than three years. A herd bull must produce sons better than he is within that time span or the herd is not progressing genetically, Cooper has said. Selections of sons as successors is mainly on the basis of adjusted weaning and yearling weight per day of age. Progeny testing, they add, is too slow. "By the time data on steers could be obtained," says Holden, "we may be using sons, or even grandsons, of the sire of the steers."

The Montanans are equally rigid in maintaining their standards in their cow herds. They apply a 60-day breeding season to both cows and heifers, discarding any that fail to settle in calf within that period unless it is felt that there is a nutritional reason for the failure. Not only does this practice result in calf crops of uniform age but also it contributes to maintaining the fertility of the cow herd by eliminating the slow breeders.

If a brood cow is what she should be, says Holden, she will produce a heifer better than she is. These top performers from the heifer crop then are used to replace cows culled off at the bottom end of the herd. The two herd owners are firm believers in thus rapidly turning over the generations. "It is hard to believe," Holden has said, "how widely performance varies between the top end and the bottom end of the cow herd until their performance records are studied." Both men feel that there is still work to be done in most herds, including their own, in bringing the bottom up nearer to the top.

The cow must milk well enough to raise a well-grown-out calf and still stay in good enough condition to breed back promptly, they insist. Any defects likely to affect productivity in the cow herd are ruthlessly eliminated by culling, and both men are well aware of the desirability of maintaining acceptable conformation characteristics. Holden says: "We like to visually

*Husky youngsters were a feature of this "family portrait" taken at the Holden Hereford Ranch.*—Drovers Journal *photo.*

appraise our cattle to determine if they are functionally sound and if their over-all appearance is saleable." He also desires a little above average degree of red pigment in the skin to minimize the possibility of snow-burn and cancer-eye problems.

For their extraordinary accomplishments, Jack Cooper and Leslie Holden have been widely honored. Signal recognition in the Hereford industry, of course, came with their induction into Herefordom's Honor Gallery. Both were cited in particular for their rigid selection practices and production-testing standards which had enabled them to outstandingly increase the calf weights in their herds.

Numerous other honors also have come to these two. Both were honored in 1975 by the Beef Improvement Federation as "Seedstock Producers of the Year," the first time this recognition had been jointly given. Jack Cooper was declared "Outstanding Cattleman" by the Gallatin Beef Cattle Producers in 1975, won the "Agricultural Recognition" award of the Bozeman Chamber of Commerce in 1979, and was named "Outstanding Agriculturist" by the Bozeman Chapter of Alpha Zeta in 1980. He is a former 4-H leader. Leslie Holden is a past president of the Montana Beef Producers Association and a director of the Beef Improvement Federation. Both men also have held official posts in the Montana Beef Performance Association, and are otherwise involved in industry activities.

Sons are carrying on the cattle-raising traditions of both families. Les and Ethel Holden have two sons, both in business for themselves. Scott Holden raises registered Herefords at Absarokee, Montana, while John Holden has his own place near Valier where he is producing Polled Herefords. Mark

Cooper, son of Jack and Phyllis Cooper, is engaged with his parents in operating the Cooper Hereford Ranch.

The recognition which has come to Jack Cooper and Leslie Holden indicates how well they succeeded in fulfilling their ambition to produce practical cattle well above average in performance, which would yield profitable returns for the commercial cattlemen whom they perceived from the outset to be their main customers. That philosophy and the steadfast pursuit of it provided a solid foundation for their success, then and through the years which followed. Jack Cooper once voiced that concern in these words: "I don't like to sell a bull for breeding purposes to anyone that we don't think will do the buyer any good."

The performance records made by the Cooper and Holden entries in Montana "test" events during the 1960's not only impressed the open-minded Montana cowmen who were seeking better returns from their feeder calf crops but also began to attract attention from other purebred Hereford breeders not only in Montana but also in other states. So much so that visitors began to come from Texas, Oklahoma, Virginia and elsewhere to see the cattle that were beginning to make national news. Some of these visitors became buyers, and in their new homes these cattle attracted still other buyers. And almost before they recognized what was happening, the Montana commercial ranchers, plus some other home-state purebred breeders, found themselves competing against registered Hereford breeders from a dozen or more states for the Cooper and Holden bull output.

By the latter 1960's Holden and Cooper decided that the time had come to try the auction route in marketing the prime products of their program—the top 70 percent of their bull calves. By April, in 1969, the month in which all of their subsequent joint sales were held, and at Great Falls, Montana, where their auctions were held as long as they sold their output in joint sales, 44½ lots averaged $1,207. A year later, in a sale to which Scott Holden, son of Les Holden, also was a contributor, 54¾ lots of bulls commanded an average of $2,180, with a top bid of $16,000. The breadth of interest in the offering was indicated by the first sentence in the sale report in *The American Hereford Journal*: "A tremendous gathering of prominent cattlemen from across the U. S. showed keen interest in the 1969 bull calf production. . . ." The 1971 sale averaged $1,907 on a somewhat larger offering, with the names of some of the country's leading breeders appearing in the buyers' list. The next year the mark on 68 bulls was $2,235.

In the vernacular, the demand for Cooper and Holden Herefords "caught fire" in 1973 and continued at a torrid pace through the remainder of the '70's and into the '80's. The 1973 event resulted in an average of $4,591 with a top price of $26,500. Another huge advance was recorded in the next year's spring sale in which 59 bulls averaged $8,248. No fewer than 18 bulls in that auction commanded prices of more than $10,000 each.

The successful pattern of Holden and Cooper spring auction sales continued thereafter until finally, in the 1979 event, the Montana half-brothers sold 71 bulls, all calves or short yearlings, with a single exception, at an average of $12,887. This was a new record mark for a Hereford bull production sale. The average on the top 10 bulls was $45,950, with a Cooper entry bringing $92,500 to Pedretti Ranches at El Nido, California, and a Holden bull going at $90,000 to Harrell Hereford Ranch of Baker, Oregon.

These two had established a new record average for a female sale the previous fall–October 26, 1978, to be exact. In it they sent 68 head, representing all of their cows more than five years old, plus a few heifers, through the salering at Great Falls for an average price of $6,818. The top-selling matron went at $35,000, a mark which matched the previous all-time record price paid for a Hereford female.

It was highly appropriate that these climactic events should have marked the conclusion of the joint auction sales of Leslie Holden and Jack Cooper. As he approached the age of 68, the former decided to begin the marketing of his calves in the fall in order to ease his tasks during the frigid winters of upper Montana. In his first individual sale on October 4, 1979–and the first ever staged at his home ranch–he sold 42 spring bull calves right off their dams for an average of $6,994, "a breed record for a sale of this kind," according to *The American Hereford Journal*. He also included in the offering 24 cows and heifers which averaged $4,991.

Then it was Jack Cooper's turn to hold his first sale on the customary spring

*A typical group of cows and calves on a dryland pasture at the Cooper Hereford Ranch, a focal point for visitors bent upon herd improvement through the '70's and beyond.–Cooper photo.*

date, and that 1980 event, too, was an overwhelming success. Selling for the first time at his home place, his 44 bulls, all but one just a couple of months or so past a year old, averaged $14,920, with the top-seller going to Weldon Edwards of Clyde, Texas, and Lee Campbell of Dublin, Texas, at $180,000, hailed as a record price for a yearling bull. In this auction, Cooper also sold 28 cows and heifers at an average of $5,580, with a top bid of $14,000 being recorded in this category.

Then on October 2, 1980, only five weeks before his induction into the Honor Gallery, Les Holden staged his second individual sale. The average this time was $10,163 on 54 head, with the offering consisting entirely of 1980 bull calves with a single exception. That exception was the three-year-old herd bull, C L1 Domino 784, which brought $200,000. The top-selling cow brought a record $40,000 price. Buyer of both was Rocking Chair Ranch of Fort McKavett, Texas.

Neither Holden nor Cooper has chosen to fit show herds for exhibition at the fairs and stock shows. They are not at all opposed to the showring, but their preference has been to concentrate all of their time and effort on the operation and improvement of their breeding herds and their merchandising activities. Further, neither man is personally inclined toward a showring program. But cattle which they have sold, and the descendants of those cattle, have been exhibited with outstanding success.

For example, the Pruett-Wray Cattle Company of Sasabe, Arizona, an extensive and intensive user of Cooper-Holden bloodlines, exhibited the grand champion carload of bulls at Denver's National Western Stock Show for four consecutive years—1978, 1979, 1980 and 1981—a feat not duplicated by any other exhibitor thus far in the last half of the 20th century.

Herefords of similar descent, many of them sired by bulls acquired direct from Cooper or Holden, have been remarkably successful in individual competition, winning purple-ribbon honors at such major Register of Merit events as those at Denver, Kansas City and elsewhere. As early as 1975, six of the 10 bull-class winners at one ROM show carried Holden-Cooper bloodlines, and in another Register of Merit contest both champions, as well as seven of 10 class winners carried the same breeding. The same family has yielded top champions of both sexes in recent years at that greatest of all annual Hereford events, Denver's National Western. But to detail the full list of such winnings seems needless.

As also would be a recital of the long prices which they have brought in other sellers' auctions. Prices ranging upward from $100,000 have not been uncommon. The high point was reached in the fall of 1980 when Pruett-Wray Cattle Company sold a one-quarter interest in CH Domino 053 for $233,000, thus establishing a total valuation of $932,000, hailed as a new breed record. Purchaser of that one-quarter interest was the Ideal Breeders Cooperative, a partnership composed of Gene Meitler of Lucas, Kansas, as chairman; Rutt Herefords of Campbell, Nebraska; Tom Dean of Cherokee, Texas; Tien Herefords of Prairie View, Kansas; Oleen Cattle Company of Falun, Kansas; Jamison Herefords of Quinter, Kansas; Schlickau Herefords of Haven, Kansas; Vel-Leo Herefords of DeWitt, Nebraska, and Parcel Herefords of Coldwater, Kansas.

The question naturally arises as to whether the Line 1's made Leslie Holden and Jack Cooper famous, or, whether Jack Cooper and Leslie Holden, through their constructive adherence to it, brought fame to the Line 1 family. There is no doubt that their faithful and consistent practice of recording weights and measurements of significance in yielding performance data, and then carefully using this information to advance their program, and staying with it when others scoffed, brought them to the pinnacle of success during the decade of the 1970's.

Within a span of little more than a dozen years, few breeders in Hereford history have emerged with greater force from relative obscurity to positions of national, and even international, prominence. Bloodlines which descended from their herds have been infused into leading herds from coast to coast, and the diffusion of this concentrated genetic power yielded an immediate impact seldom seen in any major breed of livestock.

There was not the slightest doubt, as the 1980's arrived, that these two men deserved to rank in the top echelon of those who have contributed immeasurably to the Hereford breed's position in the beef-production industry.

# DONALD R. ORNDUFF

KANSAS CITY, MISSOURI

1907-

By B. C. SNIDOW,
as recalled by the subject

THE influence of some families has persisted through several generations of Hereford activity. This is a natural sequence in families in which sons and grandsons have succeeded fathers and grandfathers–and on and on–in herd operations. But how can the influence of a family be carried forward for more than a half-century after that family's personal connection with Herefords ceased to exist?

This is a fair question and it can be simply explained. An example is provided by the historic Sotham family and its influence on the career of Donald R. Ornduff, who was installed in the Hereford Heritage Hall's Honor Gallery on November 10, 1980.

The first Sotham in America, the English-born William H. Sotham, brought across the Atlantic from his native country the first group of Herefords of sufficient numbers to comprise a breeding-herd unit. His son, T. F. B. Sotham, during the latter years of the 19th century, gained wide recognition as one of the great Hereford breeders, and exponents of the breed, of his era. And, finally, this Sotham and his three sons were extensively engaged for a decade or so, mainly between 1910 and 1920, in the sale management field, handling such significant auctions as the dispersion in 1916 of the famous herd of Gudgell & Simpson at Independence, Missouri.

It was during this period that the Sotham influence, however subtly and unintended, fell upon an Iowa farm boy, Donald Ornduff. For some unexplainable reason, the name of his father, S. Ornduff, who was not a Hereford breeder and who never had a serious intention of becoming one, found its way onto the Sotham mailing list. And the uninterested father began to find in his mailbox on a rural route out of Milton, Iowa, copies of the catalogues of Hereford sales which T. F. B. Sotham and his sons were managing.

But those catalogues, with their colorful covers and pictures of good Herefords on the inside pages, proved so fascinating to the son that he still had in his files, as of 1981, the very first one received. It was issued for the sale of A. A. Berry & Son, near Cedar Rapids, Iowa, held on September 1, 1916. Never mind that the boy, then not yet nine years old, could little have comprehended the contents of that Sotham-produced catalogue. The time came when he could–and did. And it was to the credit of the Sothams that his initial interest was aroused.

Two or three years later, a nearby Hereford breeder, Jess Bell, obviously intending to encourage the father to become interested, sent him a year's subscription to *The American Hereford Journal*. While the father's interest was not aroused, the son more than 60 years later still remembers very well his excitement with the arrival of that first issue. It was that of January 15, 1919, featuring the report of a record-breaking sale which had just been held by Mousel Bros. of Cambridge, Nebraska. Don Ornduff filed that issue as he did hundreds of others which followed, as he became a subscriber on his own

*Don Ornduff, at right, then editor of* The American Hereford Journal, *visited in the judging ring of Chicago's International Live Stock Exposition on a December day in 1953 with Elwyn O. Jones, then owner of the famous Atok Herefords at Sheephouse, Hay-on-Wye, Hereford, England.–AHA photo.*

when the gift subscription expired. The cost in those times was only a dollar a year.

While yet in high school, young Ornduff took a significant step, with the idea in mind of some day working for the *Journal.* He traded a grade Hereford heifer which he had bought with money earned in working for neighboring farmers to a local store owner for a typewriter. His course away from the direction of becoming a farmer-stockman thus was set. And away from home he went for more schooling.

When he finished his college studies the depression of the late 1920s was in full swing. *The Hereford Journal,* as with everything else dealing with agriculture, was struggling. There was no opening. So he went to work for John Morrell & Co., the meat-packing firm, at Ottumwa, Iowa, but continued to maintain his contact with Hayes Walker, Sr., the founder of the *Hereford Journal.*

When word of an opening came in 1930, Ornduff did not hesitate to make a change. That his decision was the right one is evident in the fact that he stayed with the *Journal* for 40 years–a five-years longer span than anyone else ever associated with the magazine. And after retirement from full-time duties, he continued to be active as contributing editor. Thus, through five decades his dedication to the Hereford breed, to Hereford breeders and to the Hereford industry did not waver.

Starting with the *Journal* as circulation manager, and assisting the editor, John M. Hazelton, occasionally, he soon made a favorable impression, as is indicated by this somewhat prophetic reference in an article by Hayes Walker, Sr., entitled *"The Hereford Journal, Past, Present and Future,"* which was published in the issue of July 1, 1931: "Don Ornduff, a younger member of the staff, has been for many years one of the keenest students of the breed," the *Journal's* publisher wrote, "and his contributions personally and through the columns of *The Hereford Journal* will be of genuine value to our readers."

In an article on the history of the *Journal,* published in its Silver Anniversary Edition in 1935, Hayes Walker, Sr., concluded his reference to Ornduff, who by then had assumed an editorial management position after John Hazelton's resignation: "I know of few men whose knowledge of Hereford history is so nearly complete."

At length, Ornduff became a member of the board of directors of Walker Publications, Inc., in which the *Journal's* ownership was vested, until its sale early in 1961 to Hereford Publications, Inc., a subsidiary of the American Hereford Association. He then was a member of the board and vice-president and/or secretary of the latter corporation until his retirement.

Ornduff's research continued through all of these years. He sought out

*The span of Don Ornduff's connection with Hereford interests is indicated by this picture. On September 8, 1970, the day he retired as editor of the* Journal, *after exactly 40 years as a member of the staff, he stood in front of a complete set of bound volumes covering the magazine's then 60-year history. Of the 1,441 issues of the* Journal *which had been printed between the date of its beginning in 1910 until his retirement, Ornduff had contributed as a staff member to the issuance of 945 of them—more than five full shelves beginning with the volume to the right of his hand and continuing below.*—Hereford Journal *photo.*

old-time breeders and, in addition to conversations with several of them, acquired files of old livestock publications and Hereford sale catalogues dating well back into the 1880's which he studied and maintained for future reference purposes. One of these sources was Frank Gudgell, son of Charles Gudgell, who gave him a small catalogue listing the pedigrees of all the Herefords in the famous Gudgell & Simpson importation of 1881 which included Anxiety 4th. This acquisition resulted from a visit the two had at a Mousel Bros. sale in 1939. This morocco-bound booklet, of which only a few were issued by T. Rogers of Hereford, England, who assembled the cattle for shipment, probably is the rarest item in the entire field of Hereford memorabilia. Over all, it is doubtful if any file of historic Hereford significance, comparable to Ornduff's, is in existence.

On the basis of this foundation, plus day-by-day contacts with Hereford events and personalities, he wrote literally thousands and thousands of pages of material for publication in the *Journal* and in other publications in the

*Don Ornduff spoke in 1976 at the dedication of the Nita Stewart Haley Memorial Library and J. Evetts Haley History Center at Midland, Texas, especially on Haley's biography of the famous rangeman, Charles Goodnight. The library and history center almost immediately gained wide recognition as one of the country's foremost western and range industry research institutions.–Cody Davis photo.*

United States, Canada, England, Australia and South America. As recently as 1978, several years after his official retirement, he traveled to England's Herefordshire, not only to see some of that famous region's current-day Herefords but also to view the farm and farmstead where Anxiety 4th, the so-called "father of American Herefords," was bred and born. Too, he found inspiration in a visit to nearby King's Pyon to see where the breed was founded under the tutelage of Benjamin Tomkins, and saw there also the grave of Tomkins in the country churchyard.

Even though most of his time as editor of the *Journal*–in which post he succeeded Hayes Walker, Sr., in 1944–was spent on inside duties in "getting the paper out," he nevertheless attended many historic shows, including, of course, the American Royal, plus the National Western (as recently as 1981), San Francisco's Grand National, Fort Worth's Southwestern, the Chicago International for more than 30 years in succession, and a variety of state fairs and other shows.

The 1970 American Royal was dedicated to him as "The Don Ornduff Register of Merit Hereford Show," only the fourth time that such recognition had been extended in the up-to-then 72-year history of the Kansas City event. The personal contacts established on all of these occasions were invaluable in providing ideas and insights which were utilized in carrying forward his responsibilities.

Likewise, and substantially for many of the same reasons, he was at the ringside for a considerable number of Hereford auctions during the years, ranging from smaller, local sales up to million-dollar history-makers. He started early on this pattern, going to the dispersion sale of William Hutcheon at Bolckow, Missouri, on September 25, 1930, only 16 days after joining the staff, and during the years which followed he attended Hereford auctions all the way from Virginia to California.

During his 10 years as managing editor, and then 25 years as editor, Ornduff cultivated a good and close working relationship with college and industry leaders who contributed their expertise in the form of articles which proved invaluable to thousands of readers. Seldom in the history of livestock journalism, it was said, had such a program ever been carried out so extensively.

Twice-a-month issues during this entire period not only kept the country's Hereford news up to date, but also carried these discussions of matters of breed and industry significance to such an extent that the *Journal's* issues, month by month over substantial periods, totaled more than 550 pages, a monthly figure not matched before or afterward. Nor did these include a Herd Bull Edition, which individually ranged upward to almost 1,100 pages in this same era. All of this news, feature and editorial material, Hereford advocacy and practical information—a distillation, in essence, of his years of study, background and activity in the field—was published under Ornduff's guidance.

It is worthy to note that although considerably involved with Hereford history, Ornduff also constantly sought to look to the future. This is evident both in the articles for which he arranged with a wide range of experts, and also in the editorial comment which he wrote regularly after the death of Hayes Walker, Sr. As far back as the earlier 1950's he became a firm exponent of formal production-testing procedures when such programs and the attendant record-keeping had found acceptance with no more than a few breeders.

It seemed inevitable, with his long-held interest in the breed and its background, that he ultimately would write a definitive study of the history of Herefords in the United States. He did, during the 1950's: the 500-page volume entitled, *The Hereford in America*. Within a few months the first edition was completely sold out, as were two subsequent editions which carried the story of American Herefords not only to every state in the U.S.A. but also into every country around the world where Herefords are in significant production.

Ornduff continued to write after his retirement from full-time *Journal* responsibilities late in 1970. Numerous articles written by him appeared not only in the *Journal* but also in such other publications as England's *Hereford Breed Journal*, the National Cowboy Hall of Fame and Western Heritage Center's *Persimmon Hill, The Drovers Journal, The Cattleman* and *The Record Stockman.*

He also wrote a book about the famous Hereford personality, Capt. Dan D. Casement, entitled *Casement of Juniata*. He wrote the extensive Historical Overview of *Bell Ranch as I Knew It*, by George F. Ellis, which in 1974 won the National Cowboy Hall of Fame's Western Heritage Award as that year's "Outstanding Western Non-Fiction Book." For that he received a replica of the bronze statuette entitled "The Wrangler," the original of which was done by Montana's great cowboy artist, Charles M. Russell. He has been associated, too, with the production of other books, either as contributor or coordinator.

This, then, covers the highlights in Don Ornduff's career in association with Herefords, but not his numerous other activities and interests. Of him, Paul Swaffar, long a close observer of Ornduff's work, and whose 13-year tenure as executive secretary of the American Hereford Association was surpassed by that of only one of his predecessors, wrote 10 years after Ornduff's retirement: "Nobody in the breed's history has done more to promote and soundly advance Herefords." Don Ornduff's unassuming nature would not permit that comment to go unchallenged, but it stands as the opinion of one close and keen observer of Hereford progress.

In looking back again to the beginning, a line or two written by Alexander Pope, noted English essayist of the 1700's, surely is applicable to Don Ornduff. It reads: "As the twig is bent the tree's inclined." As a mere boy, a twig as it were, he leaned toward Herefords. And, throughout his adult lifetime he was firmly inclined toward this great breed.

In later years, he commented that it was a lifetime of activity with no regrets. He so enjoyed doing what he did, he said, that it scarcely seemed to be work at all. Except perhaps when the hour approached for the paper to "go to press," with a week's work yet to be done in about a day-and-a-half!

*Entrance to the Hereford Heritage Hall from an* AHJ *photo by Maurice Gibbens. The Hall is at the east of the top floor of the American Hereford Association's headquarters building.*

## 1978 Inductees

**KIRKLAND B. ARMOUR**
Kansas City, Missouri

**J. S. BRIDWELL**
Wichita Falls, Texas

**DAN D. CASEMENT**
Manhattan, Kansas

**GEORGE CHANDLER**
Baker, Oregon

**HERBERT CHANDLER**
Baker, Oregon

**HENRY CLAY**
Lexington, Kentucky

**C. M. CULBERTSON**
Chicago and Newman, Illinois

**W. H. CURTICE**
Eminence, Kentucky

**ADAMS EARL**
Lafayette, Indiana

**JESSE ENGLE**
Sheridan, Missouri

**OTTO FULSCHER**
Holyoke, Colorado

**CHARLES GOODNIGHT**
Paloduro and Goodnight, Texas

**Charles Gudgell**
Independence, Missouri

**Overton Harris**
Harris, Missouri

**Robert H. Hazlett**
El Dorado, Kansas

**R. J. Kinzer**
Kansas City, Missouri

**Roy R. Largent**
Merkel, Texas

**Robert W. Lazear**
Cheyenne, Wyoming

**Murdo MacKenzie**
Denver, Colorado

**James M. McClelland**
Sulphur, Oklahoma

**T. L. Miller**
Beecher, Illinois

**Albert K. Mitchell**
Albert, New Mexico

**Earl H. Monahan**
Hyannis, Nebraska

**Henry L. Mousel**
Cambridge, Nebraska

**Robert D. Mousel**
Cambridge, Nebraska

**Thomas A. Simpson**
Independence, Missouri

**C. C. Slaughter**
Dallas, Texas

**T. F. B. Sotham**
Chillicothe, Missouri

**William H. Sotham**
Albany and Black Rock, New York

**Charles B. Stuart**
Lafayette, Indiana

**Alexander H. Swan**
Cheyenne, Wyoming

**E. H. Taylor, Jr.**
Versailles, Kentucky

**Arthur W. Thompson**
Lincoln, Nebraska

**Roy J. Turner**
Sulphur, Oklahoma

**John J. Vanier**
Salina, Kansas

**William S. Van Natta**
Fowler, Indiana

**Hayes Walker, Sr.**
Kansas City, Missouri

**Arthur D. Weber**
Manhattan, Kansas

*The 49 plaques pictured here constituted the Honor Gallery of the Hereford Heritage Hall by the third year after its inception. The sign above them reads: "Dedicated to the people who have made Herefords the number one breed in the beef cattle world. Those in this gallery led the way in developing America's rich Hereford heritage."*—AHJ *photo by Bob Day.*

## 1979 Inductees

**FRED C. DEBERARD**
Kremmling, Colorado

**CHARLES DESCHEEMAEKER**
Boyd, Montana

**CHARLES H. HARRIS**
Fort Worth, Texas

**GEORGE F. MORGAN**
Newman, Illinois, and Cheyenne, Wyoming

**FRED REPPERT**
Decatur, Indiana

The Heritage Hall Committee for 1979 was composed of Max Fulscher, Holyoke, Colorado; Burke Healey, Davis, Oklahoma; Charles Neblett, Stephenville, Texas; Donald R. Ornduff, Kansas City, Missouri; B. C. Snidow, Kansas City, Missouri, and P. H. White, Dyersburg, Tennessee.

## 1980 Inductees

**J. P. CALLIHAM**
Conway, Texas

**FARRINGTON R. CARPENTER**
Hayden, Colorado

**JACK COOPER**
Willow Creek, Montana

**LESLIE HOLDEN**
Valier, Montana

**DONALD R. ORNDUFF**
Kansas City, Missouri

**LESTER WIESE**
Manning, Iowa

The Heritage Hall Committee for 1980 was composed of Max Fulscher, Holyoke, Colorado; Burke Healey, Davis, Oklahoma; Charles Neblett, Stephenville, Texas; B. C. Snidow, Kansas City, Missouri; P. H. White, Dyersburg, Tennessee, and Gene Wiese, Manning, Iowa.

# The Founding Committee

*Selected by the AHA Board of Directors to develop the format and guidelines for future development of the Hereford Heritage Hall was the committee pictured above: (Left to right) Donald R. Ornduff, Kansas City, Mo., former editor of* The American Hereford Journal *and author of* The Hereford in America; *Walter Bones, Hereford breeder and owner of Bones Hereford Ranch, Parker, S.D.; B. C. Snidow, Assistant Secretary, American Hereford Association, Kansas City, Mo.; Charles Neblett, Hereford breeder of Stephenville, Texas; Max Fulscher, Hereford breeder of Holyoke, Colo., and Burke Healey, co-owner of Healey Bros. Flying L Ranch, Davis, Okla. This committee selected the 1978 inductees, as listed on pages 278 and 279.*

## Gallery Selection Procedure–

1. Suggestions or nominations of future Honor Gallery inductees will be welcomed by the Heritage Hall committee from any source.
2. The maximum number of inductions in any year after 1978 will be limited to no more than five (5) except possibly in 1981 which will mark the 100th year of the Association's existence when the number may be increased to ten (10). In all cases the number will be determined by the committee.
3. Each nominee must have a seconding motion from a committee member to be considered in the final selection.
4. Seconded nominations will be submitted to the committee as one roster.
5. In a run-off procedure the committee will select the year's prospective honorees.
6. Final selection will require the approval of at least five of the six-member committee.